COLLABORATIVE MODELING AND DECISION-MAKING FOR COMPLEX ENERGY SYSTEMS

COLLABORATIVE MODELING AND DECISION-MAKING FOR COMPLEX ENERGY SYSTEMS

Ali Mostashari

Stevens Institute of Technology, USA

World Scientific

NEW JERSEY · LONDON · SINGAPORE · BEIJING · SHANGHAI · HONG KONG · TAIPEI · CHENNAI

Published by

World Scientific Publishing Co. Pte. Ltd.

5 Toh Tuck Link, Singapore 596224

USA office: 27 Warren Street, Suite 401-402, Hackensack, NJ 07601

UK office: 57 Shelton Street, Covent Garden, London WC2H 9HE

British Library Cataloguing-in-Publication Data
A catalogue record for this book is available from the British Library.

ISBN-13 978-981-4335-19-5
ISBN-10 981-4335-19-3

Typeset by Stallion Press
Email: enquiries@stallionpress.com

Printed in Singapore by World Scientific Printers.

Contents

Chapter 1   Introduction   1

At the Outset   1

1.1   Motivation and Context: Engineering Systems   4
      and Stakeholder Involvement
1.2   Significance of the Book Topic   8
1.3   Towards a Better Energy Systems   12
      Decision-making Process
1.4   Approach   12
1.5   The Importance of Systems Representation   16
1.6   Chapter Summary   20

Chapter 2   Systems Analysis   23

2.1   What are Engineering Systems?   23
2.2   Engineering Systems Analysis Methodologies   31
2.3   Critique of Engineering Systems Methodologies   45
2.4   Chapter Summary   48

**Chapter 3   The Role of Expert Analysis in Complex   51
           Systems Decisions**

3.1   The Role of Technical Expertise in Engineering   52
      Systems Policy-making
3.2   Challenges for Effective Use of Science   54
      in Engineering Systems Policy

3.3   Perceived Technical Expert Bias and Scientific   56
       Advocacy
3.4   Communicating Science and Scientific   62
       Uncertainty
3.5   Interactions Among Stakeholders, Decision-   66
       makers and Technical Experts/Experts
3.6   System Representations and System Models   69
       as Boundary Objects in Science-Intensive
       Disputes
3.7   Obstacles to Increasing the Role of Expertise   72
       in Decision-making
3.8   Chapter Summary   74

**Chapter 4   Systems Representation and   77
              Decision-making**

4.1   Representations and the Abstraction of Reality   78
4.2   Internal Representation: Mental Maps   79
4.3   External Representation: Words and Imagery   80
4.4   Representations, Beliefs and Value Systems   82
4.5   Representation and Bias   83
4.6   Engineering Systems Representation   85
4.7   Experiments in Engineering Systems   89
       Representation
4.8   Stakeholders, Conflict and Systems   95
       Representation
4.9   Challenges of Involving Stakeholders in   96
       Engineering Systems Representation
4.10   Designing an Effective Stakeholder-Assisted   99
       Representation Process
4.11   Limitations of System Representations as a   101
       Basis for Collaborative Processes
4.12   Chapter Summary   102

Chapter 5   Stakeholder-Assisted Modeling   **105**
and Policy Design

5.1   The Stakeholder-Assisted Modeling and   105
      Policy Design Process
5.2   Outline of the SAM-PD Process   108
5.3   Problem Identification and Process Preparation   109
      Stage
5.4   Stakeholder Assessment Stage   114
5.5   Extracting Contextual Knowledge from   122
      Stakeholder Statements
5.6   System Representation, Evaluation   127
      and Policy Design Stage
5.7   Consensus-seeking Negotiation   142
5.8   Process Effectiveness and Validity Assessment   147
      Through Peer Review
5.9   Implementation and Post-implementation Stage   151
      (CLIOS Steps 10–12)
5.10   Chapter Summary   154

Chapter 6   The Cape Wind Offshore Wind   **157**
Energy Project

6.1   Project Timeline   158
6.2   Legal Context for Offshore Wind Energy   159
      Development in Massachusetts
6.3   Project Overview   160
6.4   The Environmental Impact Assessment Process   162
6.5   Public Reaction to Cape Wind   170
6.6   Stakeholder Involvement in the Cape   173
      Wind Project
6.7   Major Sources of Dispute in the DEIS   178
6.8   Chapter Summary   180

Chapter 7   Stakeholder-Assisted Modeling   183
of Cape Wind

7.1   Applying the SAM-PD Process to the Cape   183
Wind Project
7.2   Problem Identification and Process Preparation   186
7.3   Stakeholder Conflict Assessment   189
7.4   Problem Refinement and System Definition   213
7.5   Initial Stakeholder-Assisted Representation   214
7.6   Stakeholder-Refined System Representation   241
7.7   Workshop Dynamics and Results   250
7.8   Stakeholder Feedback Survey   259
7.9   Additional Feedback from the Stakeholder   263
Survey
7.10   Comparing the Refined Stakeholder-Assisted   265
Representation with the U.S. Army Corps
of Engineers Scoping Document
7.11   Chapter Summary   278

Chapter 8   Learning from Cape Wind   279

8.1   SAM-PD Process Preparation   279
8.2   Collaborative Process Dynamics   284
8.3   System Representation as a Basis for   285
Collaborative Process
8.4   Compatibility of SAM-PD with Current   289
Permitting Process
8.5   Conclusion   291

Index   301

Chapter 1

Introduction

At the Outset

The central idea of this book is that complex energy planning processes are improved through a collaborative modeling and decision-making approach. It argues that *effective* stakeholder involvement in the conceptualization, design, implementation and management of complex engineering systems,[1] such as energy systems, are an essential part of effective decision-making. The emphasis on "effective" refers to the fact that not all stakeholder involvement results in improved decision-making.

An interesting example of the shortcomings of current energy systems decision-making processes is the Cape Wind Offshore Wind Energy Project in Nantucket Sound, Massachusetts. Proposed in 1999, it was just approved in April 2010 by the U.S. Federal Government, and lawsuits are still expected to slow its implementation. It is a classic case where "public participation" in the form of public hearings becomes a battlefield between different positions on a sociotechnical system. The result is conflict, politicization of

---

[1] Here, we define engineering systems as complex large-scale technical systems embedded within a sometimes even more complex social setting, with uncertain and often emergent long-term social, economic, environmental impacts.

the decision-making process, distrust in the ability of the decision-maker to account for public interest and the undermining of the technical and engineering analysis process. The following is a short narrative on the highlights of an actual public hearing for the Cape Wind project held in Cambridge, Massachusetts, in late 2004.

The Public Hearing Spectacle

Cambridge, MA, December 16, 2004. Third Draft Environmental Impact Statement (DEIS) Hearing held by the U.S. Army Corps of Engineers, New England District, on the Cape Wind Offshore Wind Energy Project, a proposal to build 130 wind turbines offshore in Nantucket Sound.

With the hearing scheduled for 7:00 p.m., more than 700 individuals have lined up since 5:00 p.m. to register for the two-minute slot alotted for individual comments. Small and large signs supporting or denouncing the Cape Wind project float around the corridors. A group of proponent activists dressed in yachters' clothes (portraying the opponents of the project), and calling themselves "Yachters Against Windmills Now (Y.A.W.N.)", roam the corridors chanting, "Cape Wind makes our blue blood boil…Let's get our energy from Middle East oil!..Walter Cronkite, stick to the news…We'll spend millions to protect our views!" Carrying signs that said "Global Warming: A Longer Yachting Season", the activists were met with hostile remarks by the opponents of the project carrying "S.O.S" (Save our Sound) and "No Steel Forest in Nantucket Sound" signs.

Both opponents and proponents have set up information booths in which they distribute "facts on the project" to the

(Continued)

(Continued)

public. Fewer than 1 in 10 participants choose not to wear any signs for or against the project. With the hearing beginning, one by one, members of the public start to speak on their support or opposition of the project, with many giving emotional testimonies. The comments are met with applause from parts of the audience. With the time for every comment elapsing, loud shouts of "Time! Time!" erupts from the part of the audience opposed to the speaker's position.

Substantial comments on the DEIS from independent experts are also limited to two-minute slots, although they can later submit their comments via email. One of the participants, a proponent of the Cape Wind project, decides to sing a song to the tune of the 1960s "Blowin' in the Wind" within the two-minute slot. "How many tonnes of CO_2 does it take…before we go back to the wind…how many soldiers have to die in Iraq…before you issue a permit…the answer my friend is blowin' in the wind…the answer is blowin' in the wind."

Another proponent, who flew all the way from Washington, DC, comments, "You cannot NIMBY anywhere, anytime, and expect to have electricity everywhere, all the time. I ask the opponents to accept their fair share of the burden of energy projects. Minority communities have accepted more than their fair share of pollution."

A resident of Hyannis, whose cousin died fighting in Iraq, blasted wind farm supporters for using the war in Iraq as an example of why a power source like the Nantucket Sound wind farm is needed. "Stop selling out our soldiers for a wind factory. My cousin didn't die to make Jim Gordon [the Cape Wind CEO] rich", she said and went back to her seat crying, while a proponent of the project chuckled, "Quite a Broadway performance. You have talent, lady."

1.1 Motivation and Context: Engineering Systems and Stakeholder Involvement

Those of us living in democratic societies are often reminded that our voices will be heard in one form or another when it comes to decision-making that affects our lives. Of course this voice is often not direct, and not everyone's voice is in reality represented. George Orwell's "all animals are equal but some animals are more equal than others"[2] may have been intended for a different system, but it can be used to describe the more prominent role of interest groups in shaping public policy. Hence the idea of public participation goes beyond the election of public officials, and refers to the more direct role of the public in general and stakeholders, in particular, in influencing decisions that affect them. The term "stakeholders" in this context is defined as the subset of the public at large that includes all those who have an active stake or professional interest in the system. This includes policymakers, private sector actors, non-profit organizations, citizen groups, financial institutions, scientific advisory groups, independent experts and government advisory organizations that are crucial to the success of any negotiated policies.[3]

The realization that stakeholder involvement is necessary is not recent, but the awareness of its importance has made

---

[2] George Orwell, *Animal Farm*, (New York: Harcourt Publishers, 1945).

[3] The line that separates stakeholders from the rest of the public is arbitrary and subjective. Essentially since not all 6 billion human beings living on the planet can be involved in every decision, we reduce the target audience by limiting it to those who will be more directly affected or more directly interested in the decision, even if they do not express an open interest. Practically, however, stakeholder involvement is limited to those stakeholders who are at least somehow interested to be involved, and those who have resources that allow them to partake in a decision-making process in any way.

its way into the mainstream just over the past four decades. Still, four decades seems to be ample time for stakeholder involvement to be well integrated into public decision-making practices. Particularly with regards to science- and technology-intensive decision-making processes, such as decisions made for engineering systems, the track record of success in such integration has been mixed. Despite the existence of a number of different official and regulatory channels in most advanced industrial countries, public participation in science- and technology-intensive decision-making processes has been ad hoc and often ineffective, leaving both the public and experts wary and suspicious of such processes. Stakeholder involvement has often been perceived by stakeholders as an effort by decision-makers to give the illusion that public concerns are taken into consideration, while in fact they have not. Decision-makers have made many efforts to dispel such perceptions by making increased efforts at reaching out to the public, again with mixed results. Success has been more tangible at local decision-making levels, and relatively rare on large-scale regional or national projects.

From a rational perspective, all stakeholders should support stakeholder involvement. Decision-makers are interested in making decisions that address existing problems, while keeping their constituents happy. Technical experts and experts would like their expertise to be used more effectively in decision-making processes, and not be overshadowed by politicization of such decisions. And citizens wish to have a share in decisions that affect their lives. There is ample evidence that when done right, stakeholder participation has resulted in decisions that are seen by decision-makers, experts and the concerned public as "good" decisions. We will look at these cases in later chapters. Surprisingly, the mere mention of stakeholder processes for a particular decision creates dismay among many decision-makers, experts and often even the public itself.

Why is there so much resistance and hesitation when it comes to stakeholder involvement, if its intrinsic merits are broadly accepted? One possible answer is that many of the current approaches to stakeholder involvement are inadequate and either fail at producing agreements, or fail at creating technically sound solutions. In making decisions for complex engineering systems, the challenge seems to be to design a process that can bring together different stakeholders with conflicting positions and ideologies, varying technical backgrounds and voices to the table in an effective manner, so that the decision can be scientifically and technically sound, socially "wise" and publicly acceptable.

Traditionally, decision-makers have only involved experts directly in the decision-making process for complex engineering systems, and even that on a limited basis. Environmental impact assessments and management and operation strategies for engineering systems such as transportation and energy systems are good examples of cases where decision-makers regularly use expert opinion as the main basis for their decisions. What decision-makers often expect from scientific and technical experts is a clear analysis of the behavior of the system, and the impact of different alternatives. These expectations however are often unrealistic. From a technical and scientific perspective, there is often large uncertainty in the relationship between the different components of the system, and more so when predicting how a system would behave in the long term. Furthermore, the selective inclusion of some experts and the exclusion of others by decision-makers can also exacerbate the problem, by creating an atmosphere of adversarial science, where one expert is pitted against another like gladiators in an arena. However, in the arena of technical expert-technical expert battles there are rarely victors.

The expert analysis of engineering systems is further complicated by *evaluative complexity* (Mostashari and Sussman, 2009). This type of complexity refers to the fact that such systems are nearly always embedded in a complex institutional and social system, the behavior of which is also not intuitive, and even harder to analyze. Engineering systems are characterized by the interactions of technology and institutions within a set of social and ecological systems. While technical experts are adept at analyzing the technological components, most are not experts in social and institutional issues. By emphasizing technical feasibility and optimality in the first stages of policy design process, the social and institutional feasibility of recommendations are often neglected, resulting in potentially inadequate policies for the system as a whole. Such a process fails to take into consideration stakeholders' concerns and interests as well as their local and experiential expertise. As a consequence, recommendations resulting from such a process often encounter resistance among stakeholders who have little or no understanding of the underlying logic of the recommendations. According to Vennix, Verburgh and Gubbels (1990),

> Even in the modern age of science and industrialization social policy decisions are based on incompletely-communicated mental models. The assumptions and reasoning behind a decision are not really examinable, even to the decider. The logic, if there is any, leading to a social policy is unclear to most people affected by the policy.

The preceding discussion would suggest that a better process design, that is actively designed to address the above challenges, might offer improvements to current stakeholder processes. In fact, it is the purpose of this book to propose a decision-making process in which stakeholder participation is not an add-on, but an integrated part of the decision-making process, from problem definition to system representation, design and decision-making.

1.2   Significance of the Book Topic

This book proposes the engagement of stakeholders in the modeling and policy design for complex energy systems.

There are many important reasons why an integrated stakeholder participation process is central to decision-making for energy systems. Stakeholder participation can result in benefits such as incorporation of public values into decisions, increasing the substantive quality of decisions, resolving conflict among competing interests, building trust in institutions and policymakers and educating and informing the public (Beierle, 1999). In the following paragraphs, we will look at the importance of stakeholder participation from a variety of perspectives.

A) *Ethics*: Many people agree that stakeholder participation in the public decision-making process is an inherent right of citizens in a democratic society. Decision-making on engineering systems is no exception, given the broad impacts such systems have on society at large. While not everyone affected can be involved, stakeholder participation can strike a balance between direct democracy and representative democracy.

B) *Policy implementation*: Complex large-scale engineering systems are characterized by high scientific uncertainty as well as high societal stake. When dealing with such systems, many decision-makers have drawn on technical experts and experts for advice on policy issues. However, due to lack of direct interaction between experts and stakeholders, the recommendations fail to take stakeholder interests into account.

This can result in the loss of effectiveness of the recommendations. But most often it leads to an increase in societal conflict on the issue.

C) *Adaptive management*: One of the characteristics of complex engineering systems is emergence, or behavioral complexity that is difficult to predict at the outset due to high uncertainty in the potential long-term behavior of the system. For that reason a design strategy or decision made at one point in time may be inadequate for the system at later times. To be able to adapt to emergence, it is vital that a broad-based group of stakeholders takes responsibility for managing the system over time. This is only possible if stakeholders have been involved from the outset. Stakeholder involvement allows stakeholders to take responsibility for the long-term management of the system.

D) *Reducing the cost of conflict*: Aside from an ethical imperative for deliberative and inclusive governance, a stakeholder-assisted policy design process for engineering systems can go a long way in minimizing the costs of conflict for such systems, by providing a formal structure for stakeholders to have their interests and knowledge considered in the design process. There is a large cost of conflict for large-scale engineering projects in particular, and for society as a whole. One interesting case is the proposed Mexico City airport project in 2002. The Mexican government went forward with a proposed $2.8 billion Texcoco airport project, paying more than $60 million in engineering consulting costs, without adequate involvement of the affected population (mainly subsistence farmers) in the region. About 1,000 armed farmers, calling themselves the People's Front for the Defense of Our Land, took the construction workers and engineers hostage for many days, threatening to kill them if the government went ahead with the project. The project was withdrawn by the government of President Fox, marking his first major political defeat, and providing no option of renegotiation with the farmers. Current alternative sites would be at least 2.5–4 times more expensive, so plans for the new airport have been suspended.

While they may vary in form, such conflicts are not limited to developing countries. In the United States in 1999, opponents filed lawsuits against 42% of large-scale project permits approved through the National Environmental Protection Act (NEPA). These included new roads, watersheds, logging operations, energy facilities, telecommunication developments and industrial sites. Direct litigation costs for such projects can range from $2 to $7 million[4] per case, and can delay projects for months and sometimes for years.[5] The annual cost of disputes over such projects is in the billions of dollars, and has resulted in bitter community relations, implementation problems and delays that have increased the cost of the projects by several-fold. Similar implementation problems exist for managing existing engineering systems. For example, while transportation-related emissions in Mexico City make up 70% of total pollutant emissions, the many air quality strategies that have been designed by the relevant agencies have been largely ineffective due to stakeholder resistance towards implementation. Past experience has shown that the inclusion of stakeholders in the decision-making process can mitigate such conflicts, avoiding immense costs that can't be offset with the best engineering design.

---

[4] http://www.sacbee.com/static/archive/news/projects/environment/graphics/graphic3a.html (accessed May 2004).
[5] The Super 7 highway project in Connecticut has been under several injunctions for the past 20 years.

E) *System representation and cognitive biases*: Research on cognitive or mental maps show that the representation of a system and the framing of the problem are heavily dependent on the modelers' personal mental maps. The isolated scientific and technical analysis of an engineering system can result in a biased representation of the system, resulting in lower-quality recommendations, which do not fully address the important aspects of a problem. Therefore, having multiple perspectives on the system which shape the collective mental map of the system can improve the representation of the system from a scientific point of view. While the main purpose of scientific and technical advice is to supply inputs into the scientific and technical aspects of the overall policy design, good expert advice should also try to provide a minimum scientific or technical literacy to decision-makers so that they can correctly use that advice. One can also point to the responsibility of expert advice to improve cognitive or mental maps[6] of decision-makers and stakeholders as well as applying scientific thinking to the non-scientific parts of the important choices in the policy design (Dror, 2003).

It has been argued that interaction among stakeholders, experts and decision-makers, while not changing values and interests, can result in a more holistic view of a given issue for all involved. This will improve the chance that decisions are robust in the longer term, given that their underlying rationale is transparent to most people involved.

---

[6] Mental maps are the subjective interpretation by human beings of a given system based on limited and scattered information, which they connect to form a consistent image that enables them to interact with that system. Often, new information that does not fit into an individual's mental model of a system is unconsciously discarded by the mind.

1.3   Towards a Better Energy Systems Decision-making Process

As previously noted, the lack of stakeholder involvement can lead to technical optimality overshadowing social and institutional feasibility. According to Cahn (2000),

> The formal inclusion of stakeholder representatives, and by extension the public at large, goes far toward resolving the primary tensions between science and policy. Formally linking policy staff and technical experts with stakeholders creates an important linkage between technocrats and the public.

One of the ways to accomplish this is to initiate a stakeholder-assisted policy design and modeling process, where the experience and local understanding of issues embodied in the stakeholder can be captured within the model, to the extent that it does not undermine its scientific credibility.

Recommendations resulting from such a process have the potential of being accepted more readily, since stakeholders feel more ownership in the policy analysis and modeling process. Therefore, at later stages of the policy-making process, the model can then be used as a negotiations tool between the different stakeholders. The challenge is to create a model that sufficiently represents the complexity of the system, while still being understood by all the participants who are involved in the modeling process, and produce reasonable and useful recommendations.

1.4   Approach

The main thrust of this book is to propose a process by which experts, stakeholders and decision-makers can interact for making decisions on engineering systems in an integrated manner. The resulting framework, the stakeholder-assisted modeling and policy design (SAM-PD) process, makes use of insights from the currently dissociated literatures on systems modeling, systems thinking, negotiation and conflict resolution, scientific communication and linguistics. SAM-PD is custom designed to allow effective interactions among decision-makers, experts and stakeholders. The systems-centric, holistic approach of SAM-PD sets it apart from other stakeholder processes, which have been mainly developed within the field of negotiation and conflict resolution, with little emphasis on the engineering/scientific systems aspect of the problem.

1.4.1 *CLIOS Process*

The modeling effort in the SAM-PD process uses system dynamics as a tool to build a modeling framework based on the CLIOS Process, as proposed by Mostashari and Sussman (2009).

The CLIOS Process is an approach to fostering understanding of complex sociotechnical systems by using diagrams to highlight the interconnections of the subsystems in a complex system and their potential feedback structures. The motivation for the causal loop representation is to convey the structural relationships and direction of influence between the components within a system. In this manner, the diagram is an organizing mechanism for exploring the system's underlying structure and behavior, and then identifying options and strategies for improving the system's performance.

Within the context of this book, systems representation is the act of laying out the structure of a system and the linkages between its components, with the aim of characterizing its behavior and understanding its structure. In simpler terms, a system representation is a way to capture our knowledge of a system, its components and interconnections in its simplest form. The type of representation we are using in this research is known as a CLIOS diagram (a variation of what is also known as a causal loop diagram), which is essentially a picture containing words that describe system components and directed arrows connecting those components.

1.4.2 *System Dynamics and Causal Loop Diagrams*

The idea of non-technical stakeholders helping in the modeling process can be deemed unrealistic, if one imagines the thousands of lines of code or the hundreds of pages of spreadsheets that come to mind when thinking about a model. However an alternative approach is a visual model such as one created using system dynamics, which is a useful tool for system representation using CLIOS diagrams. Figure 1.1 shows a CLIOS diagram drawn with system dynamics software.

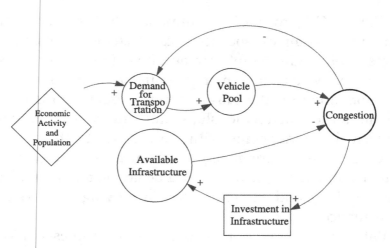

Fig. 1.1. **CLIOS diagram for a transportation system focused on the impact of infrastructure investments on congestion.**

This simple diagram shows an interesting behavior. We know that increases in population and economic activity create increased demand for transportation. If the current infrastructure availability, in this case highway or road capacity, stays constant, there will be increased congestion. Normally this prompts further investment in highway development and road construction to ease the congestion. While this indeed helps ease congestion in the short run, it will make transportation more attractive since less congestion can lead to shorter travel times and easier travel. That in turn has the effect of increasing transportation demand, which in the longer run increases congestion.

This simple diagram therefore gives us an understanding of why it is that cities like Houston, Texas, with the largest transportation infrastructures in the U.S., still experience the highest congestion levels in the nation.

System dynamics is used as a visual modeling language that can act as an accessible interface between technical modeling and stakeholders. It consists of stocks and flows and causal loops, which can explain how the different elements of a complex system are linked together. Its **qualitative representation** combined with its **quantitative output** make it a suitable tool for SAM-PD.

System dynamics also has the ability of performing extensive multi-variable sensitivity analysis. This means that if we are not certain of the inputs into the model, we can provide a range for each, and the system dynamics model will calculate all the possible combinations.

One of the major strengths of system dynamics is in simulating effects that are delayed in time. This helps us model how an event or series of events five years ago might have contributed to the status of things today, or how current policies might start to pay off in a couple of years and not immediately. The visual nature of system dynamics helps give a deeper understanding of the underlying issues through causal loops. Thus, several models with differing levels of detail can be constructed easily for different stakeholders and policy-making bodies. System dynamics can be used in conjunction with other quantitative methods such as probabilistic risk assessment, real options analysis, optimization methodologies and qualitative methodologies like scenario analysis to enable a more comprehensive quantitative analysis of engineering systems. This will be further discussed in Chapter 2.

1.4.3   *Model-Assisted Joint Fact-Finding Process*

An important part of the SAM-PD process is model-assisted joint fact-finding. The purpose of joint fact-finding is to develop shared knowledge and agreement about the system and its boundaries and important issues that ought to be considered in the scientific analysis. It is a step by which stakeholders initiate the process of gathering information, analyzing facts, and collectively making informed decisions (Ehrman and Stinson, 1999).

Joint fact-finding rests on the following main principles:

• The process of generating and using knowledge is a collaborative effort among decision-makers, independent technical experts and other stakeholders and their representative experts from all sides of the conflict.
• Information, expertise and resources will be shared among all participants.
• Participants are committed to finding a set of solutions to their conflict.

1.5   The Importance of Systems Representation

While this book proposes a decision-making process for energy systems that spans from problem definition through implementation and monitoring, its case studies mainly focus on the effect of stakeholder involvement in systems representation.

The basic rationale for the choice of systems representation as the focus of this book is as follows.

First, systems representation is one of the most important stages in any decision-making process. It is the stage at which a problem is defined, and the overall design and operation goals of an engineering system are specified. A system representation describes the boundaries of a system, its components, important performance metrics and the interconnections between the different parts of a complex system. A good representation is the foundation of later systems analysis stages, such as quantitative modeling of the representation, design and selection of alternative strategies and implementation of those strategies.

Second, furthermore, the impact of stakeholder involvement in the final decision is hard to measure in the short term; since the adequacy of collaboratively designed strategy to address a problem can only be known in hindsight, years after the implementation. However, given the importance of the system representation stage, any improvements within this stage can positively affect all subsequent stages of the process.

Third, many of the disagreements that emerge at later stages have their roots in the representation of engineering systems and the choice of performance metrics.

In this book we propose that stakeholder-assisted representation of engineering systems can result in a systems representation that is *superior* to expert representations with limited stakeholder involvement. Here, superior is defined in terms of:

- Inclusion of a plurality of views;
- Usefulness of representation as a thought expander for stakeholders;

- Usefulness of representation for suggesting strategic alternatives for improved long-term management of the system;
- Capturing effects that expert-only representation couldn't capture;
- Completeness of representation (taking into account technical, social, political and economic considerations).

An engineering system is defined through four main aspects: Its (man-made) structure and artifacts (technology, architecture, protocols, components, links, boundaries, internal complexity), its dynamics and behavior (emergence, non-linear interactions, feedback loops) and its actors/agents (conscious entities that affect or are affected by the system's intended or unintended effects on its environment). Finally, the environment it operates in also defines an engineering system.[7] Here, environment refers to the social, cultural, political, economic and legal context within which the system is operating.

A proposed taxonomy of engineering systems studies can therefore consist of:

- *Structural studies*: Research on architecture, technological artifacts, protocols and standards, networks, hierarchies, optimization and structural "ilities", etc.
- *Behavioral studies*: Research on non-linearity, dynamic or behavioral complexity, dynamic "ilities", material/energy/ information flows, dynamic programming, emergence, etc.
- *Agent/Actor system studies*: Research on decision-making under uncertainty, agent-based modeling, enterprise architecture, human-technology interactions, labor-management relations, organizational theory, lean enterprise, etc.

---

[7] One could add system goals as an important environmental constraint that defines an engineering system.

- *Policy studies*: Research on the interactions of the engineering system with its environment, including institutional context and political economy, stakeholder involvement, labor relations and social goals of engineering systems, as well as ecosystem and sustainability research.

The stakeholder-assisted modeling and policy design (SAM-PD) process research fits within the last category. It explores the impact of stakeholder involvement on engineering systems design and management and its consequences for social and environmental goals of the system. Particularly it looks at how changes in the decision-making mechanism of actors/agents can impact system structure and behavior, and how it can in turn change the effects of the engineering system on its environment.

The SAM-PD research is developed on the following engineering systems foundations:

- *Sociotechnical complexity*: One of the primary theoretical foundations of engineering systems is understanding the different kinds of complexity that affect each of the four above-mentioned aspects. Engineering systems exhibit internal (structural) complexity, behavioral (dynamic) complexity, evaluative complexity and nested complexity. Internal (structural) complexity refers to the complex structure and hierarchy of artifacts of the engineering system. Behavioral (dynamic) complexity deals with the interactions of the artifacts and the components within the engineering systems structure. Evaluative complexity refers to the fact that different stakeholders value different system performance measures differently. Nested complexity refers to the realization that complex technical systems are often nested within complex institutional structures, making the understanding of the relationship between institutions and the technical system a necessity. The SAM-PD process builds on the notions of nested complexity and evaluative complexity, in that it addresses the interactions between the institutional environment and the technical system, and involves stakeholders to jointly develop system representations and system performance goals based on a plurality of values.

- *Decision-making under uncertainty*: Another important engineering systems foundation is decision-making under uncertainty. Stakeholder involvement research focuses heavily on the fact that the communication of uncertainties and risk are essential in the decision-making process for engineering systems, particularly to those bearing the risks.

- *Engineering systems design*: Related to the above two points is the issue of engineering design. One of the foundations of engineering systems as a field is the concept of engineering systems design. The design of engineering systems is substantially different from the design of single technological artifacts, due to the existence of wide-ranging social and environmental impacts of the system, issues of emergence and the lack of a single *designer* or *architect* for systems of such scale that develop in unpredictable ways. Stakeholder involvement modifies the concept of design from an expert-centric perspective to a more inclusive process that also takes broader societal goals into consideration.

1.6   Chapter Summary

This chapter provided the foundations for the rest of the book, looking at the rationale for stakeholder involvement and providing a summary of the proposed stakeholder-assisted modeling and policy design process that can streamline stakeholder engagement in the design of policies for engineering systems. In the next chapter, we will look at engineering systems and the implications of technical complexity in engineering systems analysis.

Bibliography

Beierle, T.C. (1999) Public participation in environmental decisions: An evaluation framework using social goals. Discussion paper 99–06, Resources for the Future, Washington, DC.

Cahn, M. (2000) *Linking Science to Decision Making in Environmental Policy: Bridging the Disciplinary Gap.* Cambridge, MA: MIT Press.

Dror, Y. (2003) Science advice: Tasks, preferable features, impact assessment. *IPTS Report*, 72 (March): 13–19.

Ehrman, J.R., and Stinson, B.L. (1999) Joint fact-finding and the use of technical experts. In: Susskind, L., McKearnan, S., and Thomas-Larmer, J. (eds.), *The Consensus Building Handbook.* Thousand Oaks, CA: Sage Publications, Inc., pp. 375–399.

Mostashari, A., and Sussman, J. (2009) A framework for analysis, design and implementation of complex large-scale interconnected sociotechnological systems. Special issue on systems modeling, *International Journal of Decision Support Systems and Technologies*, 1(2): 53–68.

Orwell, G. (1945) *Animal Farm.* New York: Harcourt Publishers.

Sussman, J. (2003) Collected views on complexity in systems. Massachusetts Institute of Technology, Engineering Systems Division Working Paper Series ESD-WP-2003-01.06-ESD Internal Symposium.

Vennix, J.A.M., Verburgh, L.D., and Gubbels, J.W. (1990) Eliciting knowledge in a computer-based teaming environment. Paper presented at the Intl. System Dynamics Conference.

Chapter 2

Systems Analysis

General systems theory says that each variable in any system interacts with the other variables so thoroughly that cause and effect cannot be separated. A simple variable can be both cause and effect. Reality will not be still. And it cannot be taken apart! You cannot understand a cell, a rat, a brain structure, a family, a culture if you isolate it from its context. Relationship is everything.

— Marilyn Ferguson, *The Aquarian Conspiracy*

2.1 What are Engineering Systems?

A system can be defined as "a dynamic entity comprised of interdependent and interacting parts, characterized by inputs, processes and outputs".[8] There are many types of systems in and around us, and how a particular system is defined largely depends on where we draw its boundaries.

This book is focused on *engineering systems,* in which technologies interact with the natural and social environment in non-intuitive ways. Engineering systems

- are composed of a group of related components and sub-systems, for which the degree and nature of the relationships are not clearly understood;

---

[8] INCOSE definition.

- have large, long-lived impacts that span a wide geographical area;
- have integrated subsystems coupled through feedback loops;
- are affected by social, political and economic issues (Mostashari and Sussman, 2009).

Examples of systems that fall within this category are transportation systems, telecommunication systems, energy systems, the World Wide Web, water allocation systems, chemical industries, etc. CLIOS have wide-ranging impacts, and are characterized by different types and levels of complexity, uncertainty, risk, as well as a large number of stakeholders. In order to study and analyze an engineering system, a deep understanding of each of these aspects is necessary. In the following paragraphs, we will look at each of these aspects more closely.

2.1.1 *Complexity*

There are many definitions of complex systems, but in this context we consider a system complex when

> it is composed of a group of interrelated units (component and subsystems, to be defined), for which the degree and nature of the relationships is imperfectly known, with varying directionality, magnitude and time-scales of interactions. Its overall emergent behavior is difficult to predict, even when subsystem behavior is readily predictable (Sussman, 2003). Sussman also defines three types of complexity in systems: behavioral (also called emergence), internal (also called structural) and evaluative (Sussman, 2003).

Behavioral complexity arises when the emergent behavior of a system is difficult to predict and may be difficult to understand even after the fact. For instance, the easiest solution to traffic congestion seems to be to build new highways. New highways however cause additional traffic by attracting "latent transportation demand", due to the increased attractiveness of private autos, thus leading to more congestion in the long run.

Internal or structural complexity is a measure for the interconnectedness in the structure of a complex system, where small changes made to part of the system can result in major changes in the system output and even result in system-wide failure. A good example of this type of complexity is the side effect of chemotherapy, which in addition to destroying cancerous cells also suppresses the immune system of the body, resulting in death by infection in cancer patients.

Evaluative complexity is due to the existence of stakeholders in a complex system, and is an indication of the different normative beliefs that influence views on the system. Thus, even in the absence of the two former types of complexity, and even if one were able to model the outputs and the performance of the system, it would still be difficult to reach an agreement on what "good" system performance signifies. This type of complexity is one of the primary motivators for engaging stakeholders in systems modeling and policy design, and is an essential part of this book.

There are many different criteria for evaluating particular outcomes. Some of the social and economic valuation approaches for outcomes include

— *Utilitarian*: This criterion is one of neoclassical economics. Essentially the goal here is to maximize the sum of individual cardinal utilities: $W(x) = U_1(x) + U_2(x) + \cdots + U_n(x)$. Of course this can only function if U_i is cardinal (and if the U's are interpersonally comparable).
— *Psareto optimality*: The goal here is to reach an equilibrium that cannot be replaced by another that would increase the welfare of some people without harming others.

— *Pareto efficiency*: This occurs when one person is made better off and no one is made worse off.
— *Compensation principle*: A better-off person can compensate the worse-off person to the extent that both of them are better off (Kaldor-Hicks).
— *Social welfare function*: Here the state evaluates the outcome based on overall social welfare, taking into account distributional issues.

Which criteria are used to evaluate outcomes and how they are measured has to be determined by the consensus or overwhelming majority agreement of stakeholders. Otherwise the valuation can only be considered to be that of the experts and decision-makers alone.

Nested complexity: Finally, engineering systems exhibit nested complexity. This idea refers to the fact that a technologically complex system is often embedded or nested within a complex institutional structure. This added dimension of complexity is what makes the design and management of an engineering system a great challenge.

2.1.2 *Scale*

Large-scale systems are characterized by a large number of components, often stretching over a large geographical area or virtual nodes, and across physical, jurisdictional, disciplinary and social boundaries. Often their impacts are considered long-lived, significant and to affect a wide range of stakeholders (Mostashari and Sussman, 2009).

2.1.3 *Integration*

Subsystems within a CLIOS are connected to one another through feedback loops, often reacting with delays.

According to Sterman (2000), the existence of multiple inter-acting feedbacks makes it harder to understand the effect of one part of the system. In such a system, an institutional deci-sion may impact technological development, also impacting the environmental, economic and social aspects of a system.

2.1.4   *Environmental Interaction*

Systems may be characterized as either closed or open. A closed system is one that is self-balancing and is independent of its environment. Open systems interact with their environ-ment in order to maintain their existence. Most engineering systems are affected by the environment they operate in and, in this sense, can be considered open systems.

2.1.5   *Uncertainty and Risk in Engineering Systems*

One of the main products of complexity in a system is uncer-tainty in its initial state, its short- and long-term behavior and its outputs over time. *Webster's Dictionary* defines uncertainty as "the state of being uncertain". It further defines uncertain as "not established beyond doubt; still undecided or unknown". Uncertainty refers to a lack of factual knowledge or understanding of a subject matter, and in this case to the inability to fully characterize the structure and behavior of a system now or in the future. In analyzing complex systems, uncertainty can apply to the current state of a system and its components, as well as uncertainties on its future state and outcomes of changes to the system. Essentially there are two categories of uncertainty: reducible and irreducible. Reducible uncertainty can be reduced over time with extended observation, better tools, better measurement, etc., until it reaches a level where it can no longer be reduced. Irreducible uncertainties are inherent uncertainties due to the natural complexity of the subject matter. We can distinguish the following types of uncertainty (Walker, 2003):

Causal uncertainty: When technical experts draw causal links between different parts of the system, or between a specific input and an output, there is uncertainty in the causal link. For instance the relationship between air pollution concentration and respiratory problems is associated with causal uncertainty, given that the same air pollution concentrations can result in different levels of respiratory problems. This occurs because other, sometimes unknown factors can influence the causal link. There is also the important difference between correlation and causation, in that an existing correlation does not necessarily indicate causation. Another source of causal uncertainty is the existence of feedback loops in a system. Causal uncertainty is strongly dependent on the "mental map"[9] of the person drawing the linkages.

Measurement uncertainty: When measuring physical or social phenomena there are two types of measurement uncertainty that can arise. The first is the reliability of the measurement, and the second is its validity. Reliability refers to the repeatability of the process of measurement, or its "precision", whereas validity refers to the consistency of the measurement with other sources of data obtained in different ways, or its "accuracy". The acceptable imprecision and inaccuracy for different subject matters can be very different. For instance, the acceptable inaccuracy for a weather forecast is different from the inaccuracy of measurements for the leakage rate of a nuclear waste containment casket, given the different levels of

---

[9] A mental map is the subjective interpretation by a person or a group of people of the boundaries of a system, its linkages, its components and its behavior.

risk involved. Therefore, defining the acceptable uncertainty in measurements is a rather subjective decision.

Sampling uncertainty: It is practically impossible to measure all parts of a given system. Measurements are usually made for a limited sample, and generalized over the entire system. Such generalization beyond the sample gives rise to sampling uncertainty. Making an inference from sample data to a conclusion about the entire system creates the possibility that error will be introduced because the sample does not adequately represent that system.

Future uncertainty: The future can unfold in unpredictable ways, and future developments can impact the external environment of a system or its internal structure in ways that cannot be anticipated. This type of uncertainty is probably one of the most challenging types of uncertainty, given that there is little control over the future. However, it is possible to anticipate a wide range of future developments and simulate the effect of particular decisions or developments in a system across these potential futures. In CLIOS, the effects of new technologies often cannot be adequately determined *a priori*. Collingridge (1980) indicates that, historically, as technologies have developed and matured, negative effects have often become evident that could not have been anticipated initially (automobile emissions or nuclear power accidents and waste disposal). Despite this ignorance, a decision has to be made today.

Modeling uncertainty: Technical experts use models to predict values for some variables based on values for other variables. A model is based on assumptions about the initial state of a system (data), its structure, the processes that govern it and its output. Any of these assumptions has inherent uncertainties that can affect the results, which the model produces. The parameters and initial conditions of a model can often be more important than the relationships that govern the model in terms of the impact on the output. The "Limits to Growth" models of the 1970s show how long-range models are not capable of characterizing long-term interactions between the economy, society and the environment in an engineering system. Additionally, individual and institutional choices can make socio-economic models inherently unpredictable (Land and Schneider, 1987).

In real life, uncertainties cannot be reduced indefinitely and the reduction of uncertainty is associated with costs. Therefore an acceptable level of uncertainty for decision-making has to be determined subjectively. The subjective nature of such a determination is one of the main rationales for stakeholder participation in decision-making.

Risk is a combination of the concepts of probability (the likelihood of an outcome) and severity (the impact of an outcome). In fact, acceptable levels of uncertainty in the analysis of a system depend on acceptable levels of risk associated with that system. The concept of acceptable risk is essentially a subjective, value-based decision. While there are methodologies, such as probabilistic risk assessment, that try to provide an objective assessment of risk, it is the perception of the risk-bearing individuals, organizations or communities that determine how much risk is acceptable. While many experts focus on providing the public with probabilities of possible outcomes for a system, Sjöberg and Drottz-Sjöberg (1994) indicate that the public is more concerned about the severity than with the probability. Allan Mazur (1981) emphasizes the role of the media in affecting risk perceptions. He argues that the more people see or hear about the risks of a technology, for instance, the more concerned they will become. This effect could occur both for negative coverage as well as positive coverage.

2.2   Engineering Systems Analysis Methodologies

In this section we will look at how engineering systems have been traditionally analyzed, and what approaches can be used for engineering systems decision-making.

When analyzing an engineering system, it is necessary to look at the entire system in a holistic fashion. One of the major milestones favoring this type of systemic approach in the analysis of complex systems is *systems theory*. It was first proposed as an alternative to reductionism in the 1940s by the biologist Ludwig von Bertalanffy who published his General System Theory (Bertalanffy, 1968). He emphasized that real systems were open and that they exhibited behavioral complexity or emergence. Rather than analyzing the individual behaviors of system components in isolation, systems theory focuses on the relationship among these components as a whole and within the context of the system boundaries. According to Bertalanffy, a system can be defined by the system-environment boundary, inputs, outputs, processes, state, hierarchy, goal-directedness and its information content (Bertalanffy, 1968).

While systems theory provides the fundamental concepts for understanding a complex system, it does not provide a common methodology for analyzing such a system. In the 1960s and 1970s, systems analysis evolved as an approach to analyzing complex systems. The American Cybernetics Society defines systems analysis as

> an approach that applies systems principles to aid a decision-maker with problems of identifying, reconstructing, optimizing, and managing a system, while taking into account multiple objectives, constraints and resources. Systems analysis usually has some combination of the following: identification and re-identification of objectives, constraints, and alternative courses of action; examination

of the probable consequences of the options in terms of costs, benefits, and risks; presentation of the results in a comparative framework so that the decision maker can make an informed choice from among the options.[10]

There are many systems analysis tools and systems analysis processes that have been proposed for analyzing different aspects of complex systems.

Here we will look at systems engineering, systems dynamics and the CLIOS process as important ways to analyze CLIOS. In the following sections, we will take a look at each of these approaches.

2.2.1 *Systems Engineering*

Systems engineering is a discipline that develops and exploits structured, efficient approaches to analysis and design to solve complex engineering problems. Jenkins and youle (1971) define the following stages of a systems engineering approach to solving complex systems: systems analysis, systems design, implementation and operation.

For each of these stages, a different number of systems engineering tools and methods exist that can help analyze different aspects of the system. These methods include such elements as trade-off analysis, optimization methods (operations research), sensitivity analysis, utility theory, benefit-cost analysis, real options analysis, game theory and diverse simulation methods such as genetic algorithms or agent-based modeling.[11] At any stage of a systems engineering analysis of

---

[10] American Cybernetics Society, "Web Dictionary of Cybernetics and Systems", http://pespmc1.vub.ac.be/ASC/indexASC.html.
[11] The Institute for Systems Research, "What is systems engineering", http://www.isr.umd.edu/ISR/about/definese.html#what.

a complex system a combination of these tools and methods can be used. In the following paragraphs, we will consider each of these tools and methods and comment on their strengths and weaknesses.

Trade-off analysis: When dealing with a complex system, there are multiple values that we would like to maximize. Often, these goals and objectives can be in direct conflict with one another and maximizing one can adversely affect the other. Trade-off analysis allows us to find those outcomes in the system, which have combinations of values that are acceptable for us, and which maximize the overall value of the system as a way to deal with evaluative complexity. Multi-attribute trade-off analysis can be used for cases where there are multiple objectives in a given system. The drawback of trade-off analysis is that many benefits are not continuous in nature. For instance, in the case of offshore wind energy there is a trade-off between an open vista and cleaner energy; either there is an open vista or there is cleaner energy. The trade-off is thus not a continuous curve and cannot be well-represented using trade-off analysis.

Optimization: Optimization is the maximization or minimization of an output function from a system in the presence of various kinds of constraints. It is a way to allocate system resources such that a specific system goal is obtained in the most efficient way. Optimization uses mathematical programming (MP) techniques and simulation to achieve its goals. The most widely used MP method is linear programming, which was made into an instant success when George B. Dantzig developed the simplex method for solving linear programming problems in 1947. Other widely used MP methods are integer and mixed-integer programming, dynamic programming and different types of stochastic modeling. The

Table 2.1. A Systems Engineering Approach for Dealing with Complex Engineering Systems

System analysis	1. Recognition and formulation of the problem
	2. Organization of the project
	3. Definition of the system
	4. Definition of the wider system
	5. Definition of the objectives of the wider system
	6. Definition of the objectives of the system
	7. Definition of the overall economic criterion
	8. Information and data collection
System design	1. Forecasting
	2. Model building and simulation
	3. Optimization
	4. Control
Implementation	1. Documentation and sanction approval
	2. Construction
Operation	1. Initial operation
	2. Retrospective appraisal of the project

Source: Jenkins (1971).

choice of the methodology depends mainly on the size of the problem and the degree of uncertainty. Table 2.1 shows which methods are used for certain and uncertain conditions in the strategy evaluation and generation stages of systems analysis. Tables 2.2–2.6 identify the applications, strengths and weaknesses of various systems modeling methodologies.

Another type of optimization method is the *genetic algorithm* (GA) methodology. A genetic algorithm is an optimization algorithm based on Darwinian evolutionary mechanisms that uses a combination of random mutation, crossover and selection procedures to breed better models or solutions from an originally random starting population or sample (Wall, 1996).

Optimization methods are tools that are suitable for analyzing large-scale networks and allocation processes, but may

Table 2.2. Mathematical Programming and Simulation Modeling Methods for Engineering Systems

	Strategy evaluation	Strategy generation
Certainty	Deterministic simulation	Linear programming
	Econometric models	Network models
	Systems of simultaneous equations	Integer and mixed-integer programming
	Input-output models	Nonlinear programming
		Control theory
Uncertainty	Monte Carlo simulation	Decision theory
	Econometric models	Dynamic programming
	Stochastic processes	Inventory theory
	Queueing theory	Stochastic programming
	Reliability theory	Stochastic control theory

Statistics and subjective assessment are used in all models to determine values for parameters of the models and limits on the alternatives.
Source: Bradley, Hax and Magnanti, *Applied Mathematical Programming* (Reading, MA: Addison-Wesley, 1977).

not fit all purposes. Often when social considerations exist, the goal is not optimization, but satisfaction of all stakeholder groups involved. Also, when optimization occurs, there is no room for flexibility in the system, making the system vulnerable to changes that happen in its environment over time.

Game theory: Game theory is a branch of mathematics first developed by John von Neumann and Oskar Morgenstern in the 1940s and advanced by John Nash in the 1950s. It uses models to predict interactions between decision-making agents in a given set of conditions. It has been applied to a variety of fields such as economics, market analysis and military strategy. It can be used in a complex system where multiple agents (conscious decision-making entities) interact non-cooperatively to maximize their own benefit. The under-

Table 2.3.   Overview of Systems Modeling Methods

Methodology	Mathematical/physical foundation	Discrete/cont.	Basic components
Monte Carlo Analysis	Probability theory	Discrete, stochastic, static/dynamic	Iterations, probability distributions
Optimization	Optimization	Discrete, deterministic/stochastic, static/dynamic	Objective functions, constraints
System Dynamics	ODE, control theory	Continuous, deterministic, dynamic	Stocks, flows, delays and feedback loops
Agent-Based Modeling	Game theory, economic theory	Continuous, stochastic, dynamic	Agents, internal goals, system variables
Social Network Analysis	Graph theory	Discrete, deterministic, static	Nodes, links
Network Flow Modeling	Network theory, topology	Discrete/continuous, deterministic/stochastic, static/dynamic	Nodes, links, flows
Decision Analysis	Probability theory, risk theory	Discrete, deterministic/stochastic, static	Event/decision nodes, probabilities

Table 2.4.   Applications of Systems Modeling Methods

Methodology	Applications
Monte Carlo Analysis	Queueing and other discrete state simulation (e.g. logistics, inventory, discrete battle states
Optimization	Universal, primarily non-dynamic
System Dynamics	Any accumulation and depletion processes, systems with extensive delays/feedbacks (all types of resource management)
Agent-Based Modeling	Any heterogeneous agent system with goal-orientation, learning and behavior modification
Social Network Analysis	Any system (social/non-social) where relationship between components (architecture) is central
Network Flow Modeling	Any network system with flows
Decision Analysis	Any discrete event system where periodic decisions/events influence system outcome

lying assumption of game theory is that agents know and understand the benefits they can derive from a course of action, and that they are rational.

Agent-based modeling: Agent-based modeling is a bottom-up systems modeling approach to predicting and understanding the behavior of non-linear, multi-agent systems. An agent is a conscious decision-making element of the system that tries to maximize its local benefit. The interaction of agents in a system is a key feature of the agent-based systems. It assumes that agents communicate with each other and learn from each other. The proponents of this approach argue that human behavior in swarms (or society) within a CLIOS can only be predicted if individual behavior is considered a function of information exchange among individuals who are trying to maximize their profits (Cetin and Baydar, 2004). The main drawback of agent-based modeling approaches is that the

Table 2.5.   Strengths of Systems Modeling Methods

Methodology	Strengths
Monte Carlo Analysis	No need for exact knowledge of relationships, or variable values, well established lit., excellent sensitivity analysis, mature validation
Optimization	Powerful for finding optimal solutions in large-scale systems with numerous constraints and multiple objectives, well-established lit., excellent sensitivity analysis, mature validation
System Dynamics	Best for modeling dynamic systems with extensive feedbacks and delays, well established lit., excellent sensitivity analysis, semi-mature validation
Agent-Based Modeling	No need for exact knowledge of relationships, or variable values, best for agile, bottom-up modeling and local behavioral dynamics in heterogeneous networks
Social Network Analysis	Best for analyzing the relationship among vast numbers of components in flat or hierarchical systems.
Network Flow Modeling	Strong in modeling material and information flow in physical (and social) networks, and for modeling system "ilities"
Decision Analysis	Strong in modeling the impact of decisions/events spread over an extended period of time and for scenario analysis

initial assumptions about an individual's behavior can predetermine the aggregate system behavior, making the outcome very sensitive to the initial assumptions of the system.

Benefit-cost analysis and discounted cash flow: Benefit-cost analysis (also called cost-benefit analysis) is a methodology developed by the Army Corps of Engineers before World War II that allows decision-makers to choose projects that produce the greatest net benefit for every dollar spent. This method has been used to analyze the feasibility of complex,

Table 2.6. Weaknesses/Drawbacks of Systems Modeling Methods

Methodology	Weaknesses
Monte Carlo Analysis	Does not work for organized complexity (when extreme behaviors don't cancel each other out), not suitable for modeling dynamic systems
Optimization	Does not work well in situations where the objective function is unknown or unclear, can't handle "satisfying" situations, clumsy for dynamic systems
System Dynamics	Assumes homogeneity of similar components, doesn't allow for system memory, aggregates dynamics, can sometimes be hard to validate
Agent-Based Modeling	Immature literature, immature (and inherently challenging) validation
Social Network Analysis	Relatively static view of dynamic multi-faceted relationships, doesn't tell much about the "quality" of interactions among components, can't address delays or feedback loops easily
Network Flow Modeling	Too much focus on link-level analysis and challenging node-level analysis, reductionist approach to complex hierarchical problems
Decision Analysis	Mostly focused on decisions/events rather than systems structure, needs to be connected to other simulation and modeling methodology to be useful

large-scale projects by the public sector and the private sector. It uses the net present value (NPV) as a basis for decision-making and is used extensively to this day. The underlying assumption for this kind of analysis is that benefits and costs can be converted easily to monetary benefits and can be compared across heterogeneous projects. This can be a particularly bad assumption when dealing with social systems, where benefits are less tangible in monetary terms and evaluated differently by different stakeholders. Also, the choice of the discount rate and distributional effects are hard to capture with this methodology.

Utility theory: Utility is an economic concept that realizes that benefits of a specific good or service are not uniform across the population. Utility is a measure of the satisfaction gained from obtaining goods or services by different individuals. It can complement benefit-cost analysis by including the decision-makers' bibliography as a measure for comparison of large-scale projects. One of the problems with utility theory is that people's pbibliography can change very quickly and often there are conflicting utilities among the different decision-makers and stakeholders, making it difficult to use a single utility for a course of action or a system outcome.

Real options analysis: Real options analysis is the application of financial option pricing to real assets. Instead of the "now or never" investment options that are used in a traditional NPV analysis, real options analysis provides an opportunity but not an obligation for the decision-maker to make use of opportunities that arise under uncertain conditions. Similar to stock options, the decision-maker spends an initial investment that provides him with an opportunity to act under certain conditions to improve the value of the system he manages (Amram and Kulatilaka, 1998). A drawback of real options analysis is that it depends on a known volatility profile for any given system, something that is a far stretch for most complex systems where historical data is not necessarily predictive of future behavior.

2.2.2 System Dynamics

System dynamics is a tool for modeling complex systems with feedback that was developed by Jay Forrester at the Massachusetts Institute of Technology in the 1960s. He developed the initial ideas by applying the concepts from feedback control theory to the study of industrial systems (Forrester, 1961). One of the best-known and most controversial applications of the 1960s was urban dynamics (Forrester, 1969). It tried to explain the patterns of rapid population growth and subsequent decline that had been observed in American cities like New York, Detroit, St. Louis, Chicago, Boston and Newark. Forrester's simulation model portrayed the city as a system of interacting industries, housing and people and was one of the first systems models for a sociotechnical system. Another widely known application of system dynamics was the "Limits to Growth" study (Meadows *et al.*, 1972), which looked at the prospects for human population growth and industrial production in the global system over the next century. Using computer simulations, resource production and food supply changes in a system with growing population and consumption rates were modeled. The model predicted that societies could not grow indefinitely, and that such growth would bring the downfall of the social structure and result in catastrophic shortages of food for the world population. Given that the results of the model were highly dependent on initial assumptions as well as the designed structure, most of the predictions were not confirmed by observation in the years since, and many in the academic community have used this as evidence to discredit the value of system dynamics in modeling large-scale engineering systems. Therefore, system dynamics has in recent years shifted mostly towards solving specific problems rather than modeling entire large-scale systems. While system dynamics has made substantial progress in the past four decades, those academics not in the field still consider its merits limited, mainly because of the early large-scale experiments by Forrester and Meadows *et al.*

System dynamics uses causal loop diagrams to represent relationships and causal links between different components in a system.

In addition to qualitative representations, system dynamics also uses control theory for quantification. It uses stocks and flows along with feedback loops and delays, which can explain how the different elements of a complex system are linked together. Its qualitative representation combined with its quantitative output make it a suitable tool for modeling sociotechnical systems. In terms of quantitative capabilities, system dynamics has the ability to perform extensive multi-variable sensitivity analysis. This means that if we are not certain of the inputs into the model, we can provide a range for each, and the system dynamics model will calculate all the possible combinations and provide a range of values as the output.

One of the major strengths of system dynamics is in simulating effects that are delayed in time. This helps us model how an event or series of events five years ago might have contributed to the status of things today, or how current policies might start to pay off in a couple of years and not immediately. System dynamics emphasizes quantification of a system model as the only way to gain insights from its behavior. The CLIOS process, which uses a similar concept for representing complex systems, emphasizes qualitative insights. We will look at the CLIOS process in more detail in the upcoming section.

2.2.3 The CLIOS Process

The CLIOS process, proposed by Mostashari and Sussman (2009), is the systems analysis approach used in this book. The CLIOS process is an approach specifically designed for the study of engineering systems. The rationale for choosing this approach is its specific design that takes into consideration technical aspects of engineering systems along with their institutional side. An important distinction between CLIOS and systems engineering and system dynamics is the explicit interactions between the institutional sphere and the physical system, which enables decision-makers and stakeholders to understand the impact of their decisions and interactions on the system and provide for an opportunity of organizational improvements that allow system improvement strategies to be implemented far more effectively. Additionally, while all of the other systems approaches could benefit from stakeholder involvement, not all of them are designed to incorporate non-technical values and information. The CLIOS process however has this potential.

The CLIOS process proposes the idea of a "nested complexity" when the physical system is being "managed" by a complex organizational and policy-making system. Figure 2.1 shows the concept of nested complexity.

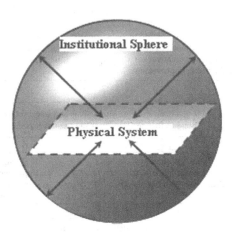

Fig. 2.1. Nested complexity: A complex physical system is nested inside a complex institutional sphere.

Source: Mostashari and Sussman (2009).

While engineering and economic models can approximate the physical system, the organizational and institutional system it is embedded in requires a more qualitative framework of analysis. Figure 2.2 outlines the 12 steps of a CLIOS process.

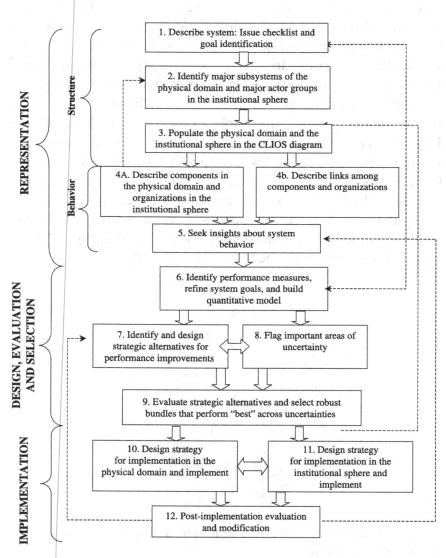

Fig. 2.2.   The twelve-step CLIOS process.

The CLIOS process has three stages:

1) representation,
2) design, evaluation and selection, and
3) implementation.

These stages cover a total of 12 steps (see Fig. 2.2). Throughout each stage and at every step, tools that are appropriate to the question at hand are used to investigate the system. The tools used for a CLIOS process can essentially be the same as those mentioned under systems engineering tools and methods, as well as those of system dynamics. A more detailed discussion on the CLIOS process can be found in Mostashari and Sussman (2009). In the following paragraphs we will take a more detailed look at the process and its steps, based on the above-mentioned paper.

In the next sections we will look at the different stages and their respective steps in more detail.

2.3   Critique of Engineering Systems Methodologies

Like methodologies in any other field, engineering systems methodologies all have their limitations and drawbacks. In this section we will discuss some of these shortcomings in more detail.

Systems engineering methodologies mainly came out of the Apollo missions of the 1960s, and were designed to enable complex engineering projects such as putting a man on the moon. Later applications were developed in military settings for large-scale projects often with unlimited funding and a hierarchical command structure. As such there is little emphasis on distributional issues, organizational impact on technological systems and issues of evaluative complexity. In the case of the Apollo missions, for example, the final goal that all decision-makers, engineers, physicists and others involved agreed on was to beat the Soviet Union in reaching the moon first. While people may have differed on how this goal should be accomplished, the underlying values were not different. Therefore traditional systems engineering approaches are not suitable to engineering systems with wide-ranging social and environmental impacts where the identification of commonly agreed performance metrics in itself is a challenge.

Optimization methods are useful tools for static systems with no social interactions. They may be useful for finding the best set of routes for airplanes in an air transportation network system or determining the optimal production capacity for manufacturing firms. They are however quite useless for engineering systems with social and environmental impacts, since it would be impossible to find a set of criteria to optimize that everyone would agree on, provided that optimal solutions exist at all. Instead engineering systems solutions are most often focused on feasibility rather than optimality.

System dynamics is a very promising systems methodology with important limitations. For one, reducing complex systems into stock and flow structures often results in the loss of important information and may lead to strategies for a system that do not address the issue in its full complexity. Additional issues are the incompatibility of social considerations with stock and flow structures. In most cases the relationship between two components is far more complicated than current system dynamics models can allow for modeling. Another criticism is the quantification of hard-to-quantify issues and values that arise in sociotechnical systems. Also important is the lack of emphasis on structures. Essentially, in a system dynamics model, structure is inferred from dynamics and not the other way around.

Most of the criticism directed by outsiders at the system dynamics community focuses on its early days in the 1960s and 1970s. The uncompromising insistence of Jay Forrester that any system could be modeled with system dynamics led to skepticism of the field by many socio technical experts, particularly when the field tried to address complex social issues by creating large-scale models and expand its application to many areas where social science scholars had emphasized the inadequacy of simplified system models that could cover the true reasons for emerging problems. However, the field has extensively evolved over time. Essentially the field has shifted away from large-scale systems modeling to problem-centered models and has mostly focused on the business community. The survival and growth of the field for over 40 years shows its intrinsic values while we bear its limitations in mind.

The *CLIOS process* is too recent and too theoretical at this point to be evaluated in terms of its strengths and weaknesses. There are many elements that can be considered an improvement over other engineering systems approaches. Particularly, by not committing to a single systems analysis tool, it enables the application of different tools in different contexts. A commonly posed criticism of processes such as the CLIOS process, which also apply (maybe even more) to the SAM-PD process, is the rational engineering structure that may not reflect the realities of complex sociotechnical systems decision-making.

With regards to systems thinking as a whole, there is always a criticism, particularly from the social sciences, that many of the abstractions used to study systems as a whole leave out a lot of important details, including important relationships, thus obscuring the real causes of issues and dynamics in a system.

2.4 Chapter Summary

In this chapter we looked at what constitutes an engineering system, and what its characteristics are. We then explored the different types of uncertainties that can arise in its analysis. We further looked at systems engineering, system dynamics and the CLIOS process as different approaches developed to analyze and improve the performance of engineering systems. The chapter also included a brief look at different tools and methods that can be used in each of the three approaches to analyze different aspects of a system. In the next chapter we will look at the role of science and technical experts and their interactions with stakeholders and the public at large in designing policies for engineering systems.

Bibliography

American Cybernetics Society. Web Dictionary of Cybernetics and Systems. http://pespmc1.vub.ac.be/ASC/indexASC.html.

Amram, M., and Kulatilaka, N. (1998) *Real Options: Managing Strategic Investment in an Uncertain World.* USA: Oxford University Press.

Bertalanffy, L.V. (1968) *General System Theory: Foundations, Developments, Applications.* New York: Braziller.

Cetin, O., and Baydar, C. (2004) Agent-based modelling for optimal trading decisions. Accenture Working Papers, February 2004.

Collingridge, D. (1980) *The Social Control Of Technology.* New York: St. Martin's Press.

Ferguson, M. (1980) *The Aquarian Conspiracy: Personal and Social Transformation in the 1980s.* Los Angeles: J.P. Tarcher.

Jenkins G.M., and Youle, P.V. (1971) *Systems Engineering: A Unifying Approach in Industry and Society.* London: C.A.Watts & Co Ltd.

Land, K.C., and Schneider, S.H. (1987) Forecasting in the social and natural sciences: An overview and statement of isomorphisms. In Land, K.C., and Schneider, S.H. (eds.), *Forecasting in the Social and Natural Sciences*. Boston: D. Reidel.

Mazur, A. (1981) Media coverage and public opinion on scientific controversies. Journal of Communication, 31(2): 106–115.

Meadows, D.H., Meadows, D.L., Randers J., and Behrens, W.W. (1972) *Limits to Growth*. New York: Potomac Associates.

Mostashari, A., and Sussman, J. (2009) A framework for analysis, design and implementation of complex large-scale interconnected sociotechnological systems. Special issue on systems modeling, *International Journal of Decision Support Systems and Technologies*, 1(2): 53–68.

Ringland, G. (1998) In *Scenario Planning: Managing for the Future,* pp. 9–27 Indianapolis: Wiley.

Sjöberg, L., and Drottz-Sjöberg, B.M. (1994) Risk perception of nuclear waste: Experts and the public. Center for Risk Research, Stockholm School of Economics, Rhizikon: Risk Research Report 16.

Sussman, J. (2003) Collected views on complexity in systems. Massachusetts Institute of Technology, Engineering Systems Division Working Paper Series ESD-WP-2003–01.06-ESD Internal Symposium.

Sterman, J.D. (2000) *Business Dynamics: Systems Thinking and Modeling for a Complex World*. Boston: McGraw-Hill/Irwin.

The Institute for Systems Research. What is systems engineering. http://www.isr.umd.edu/ISR/about/definese.html#what (last accessed May 2004).

Walker, V.R. (2003) The myth of science as a neutral arbiter for triggering precautions. *Boston College International & Comparative Law Review*, (262).

Wall, M.B (1996) A genetic algorithm for resource-constrained scheduling. PhD diss., Massachusetts Institute of Technology.

Chapter 3

The Role of Expert Analysis in Complex Systems Decisions

Whatever social or political values motivate science-intensive disputes, they often focus on technical questions that call for scientific expertise. This is tactically effective, for in all disputes broad areas of uncertainty are open to conflicting scientific interpretation. Power hinges on the ability to manipulate knowledge or to challenge the evidence that is presented to support particular policies. Both project proponents and critics use the work of "their" experts to reflect their judgments about priorities or about acceptable levels of risk. Expertise becomes one more weapon in an arsenal of political tools.

— Dorothy Nelkin, *Controversies and the Authority of Science*

Experts often provide scientific and technical advice that informs decision-makers of alternatives and their respective merits and drawbacks. Yet the role of technical experts and experts in the decision-making process is far from clear. In particular, the relationship between experts and decision-makers, between experts and society and among experts can affect the quality of decisions made for engineering systems. In this chapter we will provide an overview of the role of science, technical experts and other experts in engineering systems decision-making.

3.1 The Role of Technical Expertise in Engineering Systems Policy-making

The role of technical expertise in policy-making, specifically in the management of complex sociotechnical systems, has been increasing in the past two to three decades. According to Adler *et al.* (2000), due to increased public pressure to resolve complex and often controversial issues dealing with large-scale natural or engineered systems, policy-makers have sought better knowledge on which to base their decisions. As a result, technical experts have been more actively engaged in the creation and evaluation of knowledge used for policy purposes. In Meidinger and Antypas' (1996) comprehensive survey of the literature on the general practice of policy formulation and issues surrounding the role of science in policy, they argue that the role of science has been constantly growing in policy processes for complex systems.

However, there is increased concern that by its inability to reach out to stakeholders, science does not have a significant impact on the dynamics of the decision-making process and that the final products of the decision-making process may show little inclusion of scientific findings (Susskind, 1994).

While technical experts blame this on the politicized nature of the public policy sphere and exculpate themselves by asserting they have provided "quality science", the question remains whether scientific analysis that has little bearing on the policy process is indeed good science from a policy perspective. According to Meidinger and Antypas (1996), experience has demonstrated that the production of more scientific knowledge for policy often leads only to more questions and more controversy in areas that are already controversial. Rarely has science settled science-policy disputes, thereby raising questions about the actual role of science in the policy process.

Susskind (1994) argues for five main roles for technical experts in science-intensive disputes: trend spotting, theory building, theory testing, science communication and applied policy analysis. While his argument draws on global environmental issues as a case, these roles can be extended to many other engineering systems policy issues.

According to Dror (1999), one main function of science advice is to supply inputs into the science-related aspects of choices. However, good science advice should fulfill four additional functions regarding high-level decision-making and choice processes, such as:

(1) Providing the minimum science literacy essential for correctly using or rejecting science advice;
(2) Improving cognitive maps of decision-makers and stake-holders;
(3) Revising decision agendas;
(4) Applying scientific frames of thinking to the non-scientific dimensions of main choices.

He further argues that

Providing a balanced understanding of the scientific bases of main issues on the political agenda, including a feeling for the ambiguities and uncertainties involved, is the most important service science advice can provide the public at large. The counteracting of "magical thinking" and pseudo-science, and consequently the upgrading of public discourse as a whole, is another important task. Both policy-makers and the public at large tend to lack the necessary scientific literacy to understand many of the complex scientific issues being faced today The vast majority of senior governmental decision-makers in nearly all countries lack the minimum of science literacy required to be able to understand and use or reject science advice correctly. Furthermore, they are often unable to utilize scientific modes of thought to better comprehend complex issues. And most parts of the public in all countries are quite unable to evaluate the meanings and judge the validity of the many claims made in the name of science on topical issues, such as environmental policies, uses of biotechnology, hazardous chemicals and so on. Therefore, an important function of science advice is to provide decision-makers and the public at large with at least minimum levels of science literacy and science advice sophistication. This, in turn, requires from science advisors much more than knowledge of science (Dror, 1999).

3.2   Challenges for Effective Use of Science in Engineering Systems Policy

In their comprehensive analysis of science-intensive disputes, Adler *et al.* (2000) highlight the different obstacles that can prevent an effective decision-making process.

We have summarized and categorized their insights in Table 3.1. The issues can arise over available scientific data, expertise, the decision-making process and the proposed recommendations.

In this book, we argue that improvements in the process can help address issues on all of the four levels mentioned in the preceding paragraph. Figure 3.1 represents the traditional decision-making process for engineering systems and highlights the interactions between technical experts/experts, decision-makers and stakeholders at large. As Fig. 3.1 shows, there is a separation between the science sphere and the public policy sphere, in which the decisions are made. In many cases, the scientific and technical complexity of the natural or engineered systems in question necessitates a level of technical

Table 3.1.   Obstacles to Effective Decision-making in Science-Intensive Disputes

Issue	Potential problems
Problem definition	• Poor framing of issue
Scientific data	• Lack of access for all stakeholders
	• Inadequacy of existing data
	• Significance of presented data
	• Irrelevance of data
	• Restricted nature of data (confidentiality)
	• Inconclusiveness of data
	• Data not yet verified or not yet usable
	• Commissioned and biased data
	• Technical and scientific uncertainty
	• Outdated data
	• Data overload of stakeholders
Expertise	• Multi-disciplinary nature of problem
	• Unevenness in scientific understanding among stakeholders
	• Differential stakeholder tolerance of complexity
	• Commissioned and partisan expertise
	• Theories unsupported by actual data
	• Pseudo-expertise
	• Unrealistic expectations from technical experts
Process	• Stakeholders engaged after scientific analysis complete or near-complete
	• Stakeholder using science as a cover for other agendas
Recommendations	• Lack of economic and social feasibility of recommendations

Source: Adler *et al.* (2000).

and scientific analysis, which has traditionally resulted in the exclusion of the majority of the stakeholders from participating in the scientific analysis process.

Table 3.2 shows some of the problems with the current division between the science sphere and the public sphere

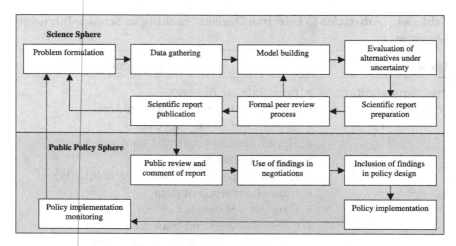

Figure 3.1. **Traditional scientific/technical analysis process for engineering systems. Dotted lines show stages that are potentially part of the process, but are often not carried out.**

that can negatively impact the role of science in decision-making and potential solutions to the problem at the different stages of knowledge generation and flow.

Based on insights from Table 3.2, one can identify important factors that affect the effectiveness of scientific advice in the policy process for engineering systems.

3.3 Perceived Technical Expert Bias and Scientific Advocacy

While there is an underlying assumption that scientific activity is by its definition objective, much debate has risen over the role of value judgments and technical experts' biases in scientific advice for public policy issues and regulatory purposes. According to Jasanoff (1990), regulatory or "mandated" science can be understood as "a hybrid activity that combines elements of scientific evidence and reasoning with large doses of social and political judgment". Values enter in

Table 3.2. Potential Problems in Different Stages of the Conventional Technical Analysis for Engineering Systems

Problems	Process stages	Possible solutions
Perceived sponsor and/or organizational bias in problem definition, choice of alternatives and findings	All stages in the scientific sphere	Independent funding for policy-related research, strong oversight on analysis and inclusion of stakeholders throughout the scientific analysis process. Elicit stakeholder inputs in choosing alternatives. Use multiple criteria for comparison, refrain from optimization
Perceived bias in model assumptions	Model building, formal peer review Process	Use of a wide range of sensible assumptions and incorporate a sensitivity analysis, agree on acceptable range of uncertainties with stakeholders. Choose wide range of reviewers and include reviewer comments and responses to critique in the final report
Uncertainty in baseline data	Data gathering, model building	Bounding some uncertainties by bounding social-eco system interaction, provision of funding for good initial data, measuring possible impact and change rather than emphasizing baseline conditions

(Continued)

Table 3.2.   (*Continued*)

Problems	Process stages	Possible solutions
Uncertainty in relationships between system components	Model building	Early stakeholder engagement and use of takeholder inputs to gain better knowledge of the system. Use of stakeholder values to bound acceptable uncertainty. Continuous reevaluation as more is known.
Uncertainty in future projection	Model building, evaluation	Use scenario analysis to bound possible future developments and draft robust strategies that perform well across different futures
Exclusion of issues of interest to stakeholders	Problem definition, evaluation of alternatives	Inclusion of stakeholders early in the scientific analysis process, starting from the problem definition
Politicization and selective use of scientific findings	Public review and comment on findings, use of findings in negotiation, inclusion of findings in policy design	Make language as unambiguous as possible and clearly explain the significance of uncertainties and the areas of the analysis they impact to avoid selective use. Promptly respond to media characterizations of the findings to prevent misrepresentation. Include stakeholders from early on in the process, make entire process transparent

(*Continued*)

Table 3.2.   (*Continued*)

Problems	Process stages	Possible solutions
Weak stakeholder understanding of the scientific process and findings	Public review and comment on findings, use of findings in negotiation, inclusion of findings in policy design	Early involvement of stakeholders in the scientific analysis. Active efforts to explain the scientific complexity and consideration of stakeholder lay knowledge in the process. Create an accessible version of the report with the important points highlighted for public understanding of the issues considered. Use an accessible report format, supported by easy-to-interpret figures and graphs. Maximize communication using new participatory techniques
Stakeholder resistance to implementation	Policy implementation	Change the process a more participatory process from the beginning and take into account stakeholder inputs and interests at all stages of the policy-making process. Take into consideration social and political feasibility in addition to technical feasibility of alternatives
No feedback between policy process and scientific analysis (open system)	All stages of the process	Change the process to a more participatory process from the beginning and take into account stakeholder inputs and interests at all stages of the policy-making process. Continuing improvement and input of science during the process. Use of scientific models in the negotiation and policy design stage

various ways, including problem definition, organization of knowledge, choice of research methodology, prioritization of critical issues and dealing with risk. Majone (1984) introduces the concept of "trans-science", which is characterized by "questions that can be stated in the language of science but are, in principle or in practice, unanswerable in purely scientific terms". He then looks at the contrasts between the U.S. and the ex-Soviet science behind the regulation of toxic elements.

While in the U.S. technical experts deem a substance to be non-toxic as long as it exists in levels that do not overload the human body's defense mechanisms or its ability to recover, Soviet science deemed a substance toxic if it evoked any physiological response at all. While the science in both cases is the same, the judgments on acceptable levels of risk are entirely subjective and relate to sociopolitical contexts in which technical experts operate (Majone, 1984). There are many sources of bias for technical experts/experts. Anderson (2000) refers to issues of educational background and dominant values and perspectives in the field of expertise, while Longino (1990) argues that technical experts can be influenced by their social, political, economic and religious values.

The subjectivity of scientific advice for public policy has not escaped stakeholders and the public at large. Limoges (1993) argues that

> confidence in the power of expertise has now vanished. For more than fifteen years, analysts of public controversies have pointed out that the involvement of technical experts in public disputes has promoted the political polarization of controversies, that expert knowledge has been almost routinely deconstructed in the course of litigations, and that expert interventions have tended to be seen as ritualistic or manipulative schemes, thus losing much of their credibility.

Being aware of the potential criticism against their biases, many "pure" and "conscientious" technical experts prefer to stick to their science. Mooney and Ehrlich (1999) indicate that in the minds of such technical experts the policy process is a linear clean process, where experts do science, advocacy groups translate the science to meet their own particular goals, and policy-makers sift the information received and balance it with general societal issues and constraints, and then make policy. This division of tasks, however, may not hold in real, controversial, science-intensive decision-making processes.

On the other hand, Weiss (1991) argues that there are good reasons for researchers to actually use their knowledge as advocacy in order to have a greater impact on policy. He argues that given the existence of values in the scientific advice process, it is more honest and productive for technical experts to explicitly state their biases. He believes that the subjective give and take involved in using research as argumentation among different technical experts may ultimately contribute to a more comprehensive picture to be formed of the issue at hand.

A drawback of this argument is that often there are subtle biases, such as disciplinary biases, educational backgrounds and institutional culture, that are not obvious to the technical experts themselves. According to Susskind (1994), adversarial science can seriously undermine the effectiveness of science in decision-making. However, he argues that there actually are legitimate sources of scientific disagreement, which can be countered by setting up representative scientific committees from all sides of the issue.

One possible solution, which is advocated in this book, is the joint expert, stakeholder and decision-maker engagement in the scientific analysis process. While stakeholders and decision-makers may not be able to perform atmospheric modeling, they are able to contribute in terms of data, resources and institutional and social feasibility. The inclusion of all science producers in a research consortium that works towards a solution for the system can help formulate more diverse, more robust solutions that are accepted across the board. Different perspectives, disciplines and backgrounds can help the group to look at the system in many more ways than any of them could individually. Additionally, the interaction provides a chance for everyone to understand the scientific analysis process, and to agree on the bounds of uncertainties acceptable for making decisions. Science-producers should not stop at the general recommendation level. They should be involved in all stages of the decision-making process, even post-implementation to review the effectiveness of policies. It is imperative that the scientific analysis team find ways to improve the chances of a recommendation being implemented by finding a "champion" for the cause without compromising scientific and technical neutrality.

3.4   Communicating Science and Scientific Uncertainty

According to Bird (2000), the lack of a general understanding of science both in society as a whole and among policy-makers is notorious. Yet even for science professionals, significant useable knowledge of scientific information outside one's own field of expertise is fairly limited. Public understanding of science and technology is clearly only one element in the development and implementation of scientific advice. Normally scientific advice develops out of the interactive communication of scientific information to policy-makers by science experts. He also points to a potentially more fundamental two-fold problem. On the one hand, policy-makers and stakeholders rarely have strong or even adequate science backgrounds and mechanisms for ensuring they obtain the science and technology information they need are limited (Brademas, 2001). On the other hand, science professionals, especially those who conduct research in academic settings, are generally not good at communicating science to those beyond their peer group (Valenti, 2000; Garrett and Bird, 2000). Furthermore, technical experts rarely acknowledge the importance of providing information to the public who fund science (Rensberger, 2000). More importantly, technical experts do not generally recognize their role in enabling the broader set of stakeholders to participate effectively in public policy decisions that depend on science. In addition, most technical experts are not trained to present scientific information to those outside their professional community, whether policy-makers or the public at large.

Arlid Underdal argues that scientific information is often in greatest demand when cause and effect relatonships are most obscure. This means that science often operates under a handicap in policy situations because it deals only with the most complex questions. While technical experts are trained to think in an uncertain world, stakeholders need to know that a recommendation will *definitely* solve the problem in question. This becomes more and more of an issue when the stakes in a problem are high, and when decisions impact many stakeholders. She further argues that technical experts are also under conditions of time pressure in which only probabilistic science with tentative conclusions can be produced (Underdal, 1989).

The decision-maker's need for certainty in the short term and the technical expert's inability to deliver it largely explains the perceived "uneasy partnership" between science and policy. One consequence of this is that many engineering

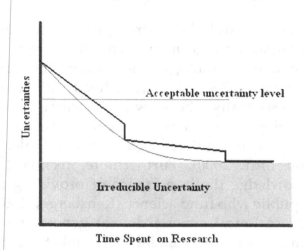

Fig. 3.2. **Proposed model for the relationship between scientific uncertainties in an engineering system and the time spent on technical research.**

systems decisions must be made in the face of fundamental uncertainty where the range of possible consequences is unknown. As an illustration, based on an interpretation of the uncertainty literature, we have proposed the conceptual representation shown in Figure 3.2. The figure shows that the scientific uncertainty for any engineering system starts at a non-infinite level and drops rapidly at first, reaching stages where the uncertainty is reduced dramatically when a threshold of knowledge on the problem is passed. The rate of uncertainty reduction however decreases over time, until it smoothes out in the long term, with irreducible uncertainty remaining for the system under study. A negative exponential that levels off at a certain stage can also approximate this trend. Exactly what levels of uncertainty are acceptable for decision-making purposes depends on the severity of potential outcomes and the resources and time available for scientific research.

3.4.1 *Science Communication Strategies*

In engineering systems decision-making, the scientific analysis used to arrive at the recommendations is often so complex that non-experts (or sometimes even outside experts) cannot understand the rationale behind the results. Therefore it has been suggested that technical experts/experts try to find ways to communicate scientific knowledge that enable decision-makers and stakeholders to make informed decisions.

Lach *et al.* (2003), asked different stakeholders, technical experts and decision-makers involved in science-intensive disputes about the importance and value of different scientific communication strategies that technical experts could use. Table 3.3 shows the results of their study. Not surprisingly, technical experts prefer to communicate the results of their studies either in academic journals or at professional conferences, while decision-makers and stakeholders prefer that technical experts directly communicate to stakeholders.

Realizing that communication is necessary is not sufficient. Most technical experts are not trained to communicate science to non-peers. As Dr. Neal Lane, head of the National Science Foundation, one of the most prestigious scientific entities in the U.S., states:

> With the exception of a few people ... we don't know how to communicate with the public. We don't understand our audience well enough — we have not taken the time to put ourselves in the shoes of a neighbor, the brother-inlaw, the person who handles our investments — to understand why it's difficult for them to hear us speak. We don't know the language and we haven't practiced it enough (cited in Hartz and Chappell, 1997).

In this regard, it is important to learn from the experience of scientific communicators in museums and popular science writers (such as Carl Sagan or Martin Rees), who are able to explain complex scientific concepts in simpler, yet still accurate terms.

3.5   Interactions Among Stakeholders, Decision-makers and Technical Experts/Experts

Based on a categorization of stakeholders by Karatzas (2001), the stakeholders in a science-intensive dispute consist of technical experts/experts who provide advice, decision-makers in government or regulatory organizations who are very often called upon to make decisions based on complex issues with social and economic implications, and stakeholders in the form of individuals or organized groups who try to influence the impact of the decision on their lives. Additionally, the role of the media is important, given the need to inform public opinion. Communicative interaction among all these groups is essential in shaping the decision for an engineering system.

Technical expert–decision-maker interactions: As discussed earlier, more and more decision-makers consult technical experts/ experts on engineering systems policy. They ask particular questions, which technical experts try to answer to the best of their ability within the timeframes required. However, according to Nelkin (1987) science-intensive disputes are never only about the technical issues involved. While the terms of the debate may be technical, the underlying issues "are a means of negotiating social relationships and of sustaining certain values, norms, and political boundaries at a time of important scientific and technological change". Questions of equity or justice arise over the allocation of resources or the distribution of economic and social costs of recommendations. Many of the recommendations that the scientific community comes up with do not consider societal or political feasibility. Once scientific results are submitted to the decision-makers, technical experts tend to leave issues of social and institutional feasibility to decision-makers, often without helping them to integrate scientific findings with such considerations. Given that the process is not really iterative, decision-makers are stuck with recommendations that may not be socially or institutionally feasible at all. Thus, they may turn their backs on science and make their decisions based on institutional interests, while giving the appearance that the science has been used as a basis. This can also lead to many scientific recommendations becoming ineffective in the actual policy-making process. The actual framing of the question, the format of the output and the consideration of social and institutional issues should be integrated into the initial scientific analysis process as important considerations.

Technical expert-stakeholder interactions: Technical experts are used to working in their own defined scientific environments, where the rules of interaction are relatively clear. They have traditionally been wary of being involved in the actual policy process, fearing that their scientific objectivity may be questioned. Therefore, they have often interacted only with decision-makers, in the form of scientific advice, and avoided direct interactions with stakeholders. This type of scientific analysis, however, which is not directly responsive to stakeholder concerns and knowledge, often leads to the ineffectiveness of science in playing its intended role as the central piece of decision-making and moves the process towards an adversarial and politicized atmosphere unlikely to produce robust and stable solutions. Stakeholders often have knowledge of the system that can be used to formulate more effective recommendations. Additionally, the engagement of stakeholders can result in their increased acceptance of the recommendations, given the feeling of process owner-ship that is developed. Most scientific analysis processes assume that once scientifically sound recommendations are out there, someone will implement them. This is often not the case, given that recommendations require actual implemen-tation plans that have to be drawn up by people who are familiar with the scientific analysis. This normally requires interdisciplinary trained experts, with good technical and sci-entific grounding as well as extensive knowledge of the policy process, who can serve as an interface between techni-cal experts and society. This highlights the importance of engaging stakeholders and decision-makers from early on, so that sound and practical implementation schedules can be drawn up.

Technical expert-technical expert interactions: Even more surprising than the lack of stakeholder engagement is the limited cooperation among technical experts and experts. As previously indicated, in the traditional science-intensive decision-making process, contesting groups have the oppor-tunity to generate competing scientific and technical knowledge to promote their positions in the decision-making process. This type of competing scientific analysis has been called "adversarial analysis" or "dueling technical experts". Ozawa (1991) cites several risks that can arise with adversarial analysis in a decision-making process. First of all she argues that the contesting groups can withhold or manipulate information to aid their respective cases, which may result in poorly informed policy decisions. Furthermore, she indicates that the knowledge claims made by any one group can become suspect in the eyes of contest-ing groups, which can delay or prevent a consensus in the process. This emerges from the fact that participants are denied a mutually acceptable foundation of knowledge upon which to negotiate policy agreements (Ozawa, 1991). Instead she proposes an alternative approach where the different groups involved in a science-intensive dispute collaborate to assemble and oversee a joint research team that aims to construct a single technical analysis of the disputed issue that will be acceptable to all of the stakeholders. Members of such a research team can be selected by representatives from all stakeholders and their progress can be continuously monitored by all those involved in the decision-making process. Resources (finances, information and equipment) are pooled to support the research and the results are accessible to all the participants (Ozawa, 1991). Pooling resources among scientific knowledge producers and working together to solve a common problem could enhance the depth and the scope of scientific knowledge to be used as the basis for an informed decision. The issue of pooled resources becomes very important in terms of available information for analysis, given that shared information can reduce many uncertainties that arise from unavailability of data. Unfortunately, most of the time this becomes an issue with financing research projects and competing grants for similar research, undermining effective cooperation.

3.6   System Representations and System Models as Boundary Objects in Science-Intensive Disputes

3.6.1   *System Models*

Scientific models are simplified representations of the important structural elements and dynamics of a system that allow us to better understand it. According to Adler *et al.* (2000), most science-intensive decision-making processes benefit from some form of modeling in order to define problems, review impacts or illustrate alternatives.

While scientific models are often thought of as descriptive and predictive of a system's behavior, they can also help improve communications in science-intensive disputes. In other words, models can serve as "boundary objects". Boland and Tenkasi (1995) define a boundary object as a visible representation of individual or community knowledge or perspectives that enables the communication of those perspectives to others in a different community. This concept supports the idea that an artifact, such as a system model that takes into account various perspectives, can mediate collaboration and serve as an interface among stakeholders, technical experts and decision-makers.

While models are useful, it is not wise to believe that it would be possible to expect a singular value generated by a model to predict a future state with absolute certainty. Stakeholders have to understand the uncertainties involved in the modeling and its assumptions. Models can help differentiate among alternatives, but cannot indicate the one true and correct answer. Therefore it is important to think of models as illustrative rather than predictive (Adler *et al.* 2000).

The use of models becomes even more confusing to the stakeholders when opposing parties bring different models to the table based on differing assumptions about inputs, interactions between variables, and outputs. The models are then staged to be in opposition to one another, when in reality they rely on different assumptions, system boundaries and initial values, and are essentially incomparable. That is why it makes more sense to develop models jointly, with various experts cooperating in its development. When such a joint development is not possible, the assumptions in the competing models have to be clear for all the experts involved.

According to Dürrenberger *et al.* (1999), good models for science-intensive decision-making processes

- should have manifest links to locally and/or personally tangible issues;
- should have a high degree of visualization and interactivity;
- should have simple structures, be transparent and have short operating/running times;
- should not be regarded as a substitute for other types of information outputs.

Also, developing models should not be an entirely expert matter. Specifically when it comes to defining the system boundaries (problem definition) and the outputs required to make an informed decision, stakeholders and decision-makers should be able to have their say.

3.6.2   *System Representations*

The decision-making process may require the use of multiple models, each dealing with a different part of the system. It is useful to have a system-wide model that combines results and models from the physical, biological, economic and social aspects of the system, and the interactions between them, to evaluate how changes in any of these aspects can affect the system as a whole. These system-wide models constitute the core focus on models in this book. Developing a system-wide model that can organize the different types of information about the system requires the presence of system modeler(s) in the decision-making process. The role of such an individual or group of individuals is to help integrate different models and types of information into a system-wide representation that will allow decision-makers, technical experts and stakeholders to make decisions on the system as a whole.

3.7 Obstacles to Increasing the Role of Expertise in Decision-making

The literature presented in this chapter is by no means comprehensive, nor can it be with an engineering systems audience in mind. The picture presented in this chapter essentially presents arguments *for* a more aggressive science role in the public sphere, but does not address the major obstacles that make such a role improbable within the current institutional structure. For one, the discussion does not emphasize the role of the embedded scientific culture and the existing reward and incentive structure that undermines the involvement of science in public decision-making. In addition to the centrality of the notion of objectivity and non-advocacy to credible science in the eyes of technical experts and experts, few incentives exist for such a change to occur. The following are some of the main issues that need to be addressed for a more active role of science/expertise in public decision-making processes, with particular emphasis on engineering systems.

Politicization of technical analysis: Most experts wish to refrain from being seen as advocates in the eyes of their peers or society. Yet, any recommendation-based technical report is often considered to strengthen the arguments of one side within an engineering system controversy. A good example of such an instance is Ted Postol's Patriot missile report that sparked a controversy at MIT, and led to serious endangerment of Postol's academic position. Postol, an MIT professor in the Science, Technology and Society program, gave testimony in front of the U.S. Congress on the ineffectiveness of the Patriot missiles during the first Gulf War. His testimony was vehemently criticized by experts hired by the manufacturer or those representing the arms industry.[12] The controversy sparked a feeling of discomfort within the MIT community, with the MIT administration looking to resolve the issue by forming an independent committee. Due to the importance of the multi-million dollar deal that was supposed to integrate Patriot missiles into the national defense budget, it took the committee five years to come to a conclusion. During this time Postol and his department came under attacks by Lockheed Martin and Raytheon, who implicitly threatened to reduce the amount of research funding made available to MIT if Postol stayed on as a faculty member.

Obstructionism and irrationality of stakeholders: Another obstacle on the side of experts is the belief that there are always stakeholders who would want to stop a process dead in its tracks no matter what the facts are. Therefore, many experts do not wish to involve themselves in a process where they feel their expertise is viewed as a bargaining chip.

Knowledge disparities and integration of stakeholder (local) knowledge: Another challenge to all the preceding discussions is the knowledge disparities among stakeholders that would make it difficult for experts to engage them when presenting their findings. Many experts are only trained to present their technical findings to audiences with similar backgrounds, and fail to interact effectively with stakeholders who may not be trained technically. Also important is the weight of

---

[12] See "The Patriot missile. Performance in the Gulf War reviewed", http://www.cdi.org/issus/bmd/Patriot.html.

stakeholder or local knowledge when compared to expert knowledge in the policy process. Experts are not comfortable with accepting stakeholder knowledge that has a different source of legitimacy than what they are used to.

Incentive structure: In the time it takes academic experts to participate in a single public policy process, they can publish several papers on more tractable issues. In addition to not being rewarded for involving themselves in such processes, they are looked down on by their peers and the academic system. Most academic institutions and technical advisory agencies do not reward interaction with stakeholders, and may in fact give preference to experts who focus on their publications.

3.8   Chapter Summary

In this chapter we looked at the role of expertise in technically complex decisions, and looked more closely at the different challenges that technical experts, decision-makers and stakeholders face in reaching informed decisions in the face of uncertainty and bias. We explored issues of communicating uncertainty and the use of models for science-intensive disputes. The discussions all pointed to the argument that while a collaborative process where decision-makers, stakeholders and technical experts all work together to reach informed decisions can overcome many of the obstacles that science-intensive disputes can pose, it cannot address all of them. Many of these obstacles have to be overcome by institutional and legal changes that go beyond a single process.

Bibliography

Adler, P.S., Barrett, R.C., Bean, M.C., Birkhoff, J.E., Ozawa, C.P., and Rudin, E.B. (2000) Managing scientific and technical information in environmental cases: Principles and practices for mediators and facilitators. Sponsored by RESOLVE, Inc., Washington, DC; US Institute for Environmental Conflict Resolution, Tucson, AZ; and Western Justice Center Foundation, Pasadena, CA.

Bransford, J., Brown, A., Cocking, R. (1999) *How People Learn: Brain, Mind, Experience, and School.* Washington, DC: National Academy Press.

Brademas, J. (2001) The provision of science advice to policymakers: A US perspective. *The IPTS Report,* 60.

Dürrenberger, G., Kastenholz, H., and Behringer, J. (1999) Integrated assessment focus groups: Bridging the gap between science and policy? *Science and Public Policy,* 26(5): 341–349.

Friedman, S.M., Dunwoody, S., Rogers, C.L., (eds.) (1999) *Communicating Uncertainty: Media Coverage of New and Controversial Science.* New Jersey, USA: Lawrence Erlbaum Associates, Inc.

Hartz, J., and Chappell, R. (1997) *Worlds Apart: How the Distance Between Science and Journalism Threatens America's Future.* Nashville: First Amendment Center.

Jasanoff, S. (1990) *The Fifth Branch: Science Advisers as Policymakers.* Cambridge, MA: Harvard University Press.

Lach, D., List, P., Steel, B., and Shindler, B. (2003) Advocacy and credibility of ecological technical experts in resource decision-making: A regional study. *BioScience,* 53(2): 170–178.

Limoges, C. (1993) Expert knowledge and decision-making in controversy contexts. *Public Understanding of Science,* 2: 417–446.

Longino, H. (1990) *Science as Social Knowledge: Values and Objectivity in Scientific Inquiry.* Princeton, New Jersey: Princeton University Press.

Majone, G. (1984) Science and trans-science in standard setting. *Science, Technology, and Human Values,* 9: 15–22.

Meidinger, E., Antypas, A. (1996) Science-intensive policy disputes: An analytical overview of the literature. Report Prepared for the People and Natural Resources Program, US

Department of Agriculture Forest Service, Pacific Northwest Research Station, Seattle, WA.

Mooney, H., and Erhlich, P. (1999) Ecologists, advocacy and public policy: Response to Wagner, F.H., Analysis and/or advocacy: What role(s) for ecologists? EcoEssay Series Number 3, National Center for Ecological Analysis and Synthesis, Santa Barbara.

Nelkin, D. (1987) Controversies and the authority of science. In Engelhardt, H.T. and Caplan, A.L. (eds.), *Scientific Controversies: Case Studies in the Resolution and Closure of Disputes in Science and Technology,* New York: Cambridge University Press, pp. 283–293.

Olson, S. (1995) *On Being a Technical Expert: Responsible Conduct in Research,* 2nd edition. Washington, DC: National Academy Press.

Ozawa, C. (1991) *Recasting Science: Consensual Procedures in Public Policy Making.* Boulder, CO: Westview Press.

Pinch, T. (2000) The golem: Uncertainty and communicating science. *Science and Engineering Ethics,* 6: 511–523.

Postol, T. (1991) The Patriot missile. Performance in the Gulf War reviewed, http://www.cdi.org/issues/bmd/Patriot.html (last accessed March 2005).

Rensberger, B. (2000) Why technical experts should cooperate with journalists. *Science and Engineering Ethics,* 6: 549–552.

Rogers, C.L. (2000) Making the audience a key participant in the science communication process. *Science and Engineering Ethics,* 6: 553–557.

Susskind, L.E. (1994) The need for a better balance between science and politics. In *Environmental Diplomacy,* New York: Oxford University press, pp. 63–78.

Valenti, J.M. (2000) Improving the technical expert/journalist conversation. *Science and Engineering Ethics,* 6: 543–548.

Chapter 4

Systems Representation and Decision-making

By relieving the mind of all unnecessary work, a good model sets it free on more advanced problems, and in effect increases the mental power of the [human] race.

— Alfred North Whitehead

In this chapter we explore the concept of systems representation, the impact of values and beliefs on representations and vice versa, how system representations shape the modeling process, and where conflict arises within representations of engineering systems. We also look at the role of modeler biases in creating system representations and the potential impacts of these biases on the design and modeling of the system. We then discuss the importance of system representations as "boundary objects" in engineering systems design, where stakeholders, decision-makers and experts can jointly strive towards a commonly agreed representation of a system of interest.

It is important to clarify some of the distinctions made in this chapter. A model can be any objective or subjective simplification of reality intended to promote an understanding of that reality. As George Box put it, "All models are wrong. We make tentative assumptions about the real world which we know are false but which we believe may be useful" (Box, 1976). Models can be conceptual (qualitative) or quantitative. Often times a conceptual model is used as a basis of quantification. In this book, we refer to the conceptual portrayal of system components and their interconnections using a diagram as systems representation, and we refer to the quantification of a system representation as systems modeling. There are many other ways of representing systems that are beyond the scope of this book. Yet it is important to realize that all types of representations are based on abstractions of reality, within the context of values and expressed as a combination of words and imagery.

4.1   Representations and the Abstraction of Reality

We often use representations to communicate our perceptions of an external reality.

By definition, representations are approximations of personal or collective abstractions of reality. Figure 4.1 shows the famous prehistoric paintings in the Lascaux caves, where the painter has tried to depict his experience of an actual hunt scene with basic representations. The representations serve as a trigger to a more or less common experience of the painter and all other individuals looking at the painting. What the prehistoric onlookers actually see is not the simple figures on the wall, but a vivid memory of an actual hunt scene they may have experienced. This aspect of representation, namely its ability to serve as an abstraction of reality, is what makes it the perfect tool for organizing knowledge. There are essentially two types of representations: internal representations (in our brain) and external representations (through which we communicate our internal presentations to other human beings).

Fig. 4.1. Lascaux cave representations of hunt scenes by prehistoric humans.

Source: http://www.students.sbc.edu/matyseksnyder04/horse%202.jpg.

4.2   Internal Representation: Mental Maps

Internal representations are the mental images or *maps* of individuals of an existing (or sometimes imaginary) external reality. The term "mental map" has been used in many contexts and was first used by Craik in *The Nature of Explanation* (Craik, 1943). Mental maps are gradually evolving heuristics that we use to categorize the knowledge we gain from the outside. They are influenced by our *values, beliefs* and *experiences*. Values reflect what we consider as acceptable or unacceptable states of a component in a system (or the system as a whole), whereas beliefs mostly influence how we perceive the relationship between the different components. Mental maps are heuristics that allow us to understand and categorize the world around us. Mental maps are created from knowledge of prior experience and feedback we have received from interacting with our environment as a whole, or a particular external object. By definition, a mental model contains the minimum information necessary for us to have a satisfactory understanding of a perceived reality. Mental maps generally have a very strong visual component. Interestingly, a change in our mental maps brought about by internalizing an external representation could also result in a change in the beliefs we hold. In its broadest sense, this process is called *learning* (Davidson and Welt, 1999).

The ability of individuals to continuously update their mental maps with new information is often called "open-mindedness". Conversely, a person with a rigid mental map may never benefit from the experiences that allow him to create a more informed mental map. Internal representations often find expression in external representations. While the internal representation of nearly all concepts is unique to each individual, it is possible for a group of individuals to agree on common external representations that reflect their internal mental map of the subject or object at hand.

4.3   External Representation: Words and Imagery

Once we have a mental map of a system, we can express the information captured in our mental map in two major ways: through words (spoken or written) and through figurative representation (images, graphs, diagrams, maps, physical structures and pictures).

Pictures and images convey multiple features of a subject/object simultaneously, while words convey information in a sequential manner. While pictures are directly stored in our brains, most of the time verbal information is interpreted using our mental maps and later referred to as contextual

Fig. 4.2. Cape Wind visual simulation (opponents).

Source: www.windstop.org.

Fig. 4.3. Cape Wind visual simulation (developer).

Source: www.capewind.org.

information. Images that we produce of an object can also reflect our values and beliefs. Figures 4.2 and 4.3 represent the simulation of the visual impact of the Cape Wind Offshore Wind Project, as produced by the opponents and the developer. While showing the wind turbines from the same vantage point from the same exact location, the images differ in contrast levels, as well as the selected zoom level, giving different impressions.

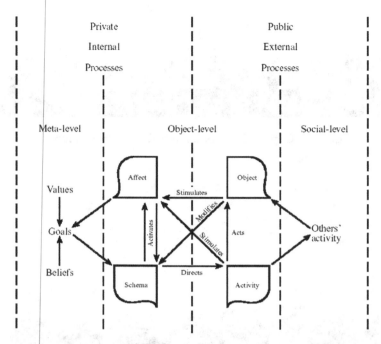

Fig. 4.4.   The relationships between values, representation and actions in human beings.

(*Source*: Marks (1999).

4.4   Representations, Beliefs and Value Systems

One of the principal roles of internal mental maps is to explore possible actions/strategies and their potential consequences before the actual realization of the action (Marks, 1999). As Figure 4.4 shows, mental processes based on mental maps and existing schemas of a problem precede actions and are impacted by external stimuli.

The ability of our minds however to conduct these thought experiments in a risk-free manner within the boundaries of our mind becomes more and more limited when the system under consideration becomes complex. In these cases, external representations can help capture many of the complexities of the system, while still retaining the values and goals of the individual. In collectively constructed external imagery, the value sets can be a combination, synthesis or co-existing set of values of the individuals involved.

4.5 Representation and Bias

The problem of having partial cognitive abilities for understanding and perceiving an external system, particularly one that is new and unknown to us, is beautifully shown in a famous poem by Jalal-ud-din Rumi (1207–1273 CE), called "Elephant in the Dark".

In Rumi's poem the elephant could be any large-scale engineering system we are confronted with today. The darkness could be taken as the cloud of complexity and uncertainty that obstructs a clear holistic view of the system. The different vantage points of the observers can be likened to the different value sets, areas of expertise and knowledge levels that stakeholders, decision-makers and experts possess when dealing with an engineering system. In fact, it can be argued that even given the same set of information and level of knowledge, system representation is heavily impacted by the mental map (how people see a system) of the group doing the representation. As we will see in experiments described later in this chapter, this becomes particularly important in engineering systems representation.

A commonly agreed representation of the elephant in Rumi's poem would indeed be possible if it were day or if, as Rumi suggested, everyone had candles and looked at the elephant from all vantage points. The candle is a metaphor for a system representation that takes different perspectives on a system and integrates them all into a coherent larger picture, so everyone can understand those parts outside their vantage point.

"Elephant in the Dark" from *Masnavi,* by the Persian poet Rumi

> *Some Hindus have an elephant to show.*
> *No one here has ever seen an elephant.*
> *They bring it at night to a dark room.*
>
> *One by one, we go in the dark and come out*
> *saying how we experience the animal.*
>
> *One of us happens to touch the trunk,*
> *"A water-pipe kind of creature is the elephant".*
>
> *Another, the ear, "A very strong, always moving*
> *back and forth, fan-animal is the elephant".*
>
> *Another, the leg, "I find it still,*
> *like a column on a temple".*
>
> *Another touches the curved back.*
> *"A leathery throne".*
>
> *Another, the cleverer, feels the tusk.*
> *"A rounded sword made of porcelain".*
> *He's proud of his description.*
>
> *Each of us touches one place*
> *and understands the whole in that way.*
>
> *The palm and the fingers feeling in the dark are*
> *how the senses explore the reality of the elephant.*

4.6   Engineering Systems Representation

System representations are ways of organizing knowledge about the system to better understand its behavior and structure, and can serve as an interface for dialogue and communication between the different individuals interested in different aspects of a system.

For an engineering system, system representations are determined by the system boundary, components and the interconnections among them. Figure 4.5 shows a typical CLIOS system diagram.

Here the boundaries of the system are drawn around the transportation demand, infrastructure availability and congestion, which are the main components under study. As explained in Chapter 1, this system representation can address the question "Can we build our way out of congestion?" but it cannot provide us with information about the impact of public transportation availability on congestion (insufficient level of detail) or the impact of transportation on air pollution (outside the boundaries of the system).

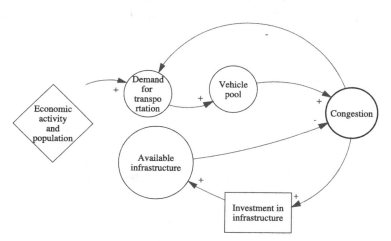

Fig. 4.5.   CLIOS diagram for a transportation system focused on the impact of infrastructure investments on congestion.

Causal loop diagrams: The above representation is a causal loop diagram (CLD). It shows the cause and effect relationship between the different components. Causal loop diagrams are used in system dynamics and the CLIOS process, making it possible to use system dynamics software to portray CLIOS diagrams. CLDs can be used to gain qualitative insight on how the different parts of a system interact. Unless quantified, a causal loop diagram cannot tell us how strong each of the linkages are or the functional form that the relationships take.

Stock and flow diagrams: Another type of engineering systems representation is a stock and flow diagram (SFD). SFD diagrams are ways of representing the structure of a system with more detailed information than CLDs. Stocks are fundamental to generating behavior in a system; flows (rates) cause stocks to change. Stock and flow diagrams are the basis for building a system dynamics simulation model. While CLDs can represent both the behavior and structure of a system, SFDs mostly focus on the behavior or dynamics, providing an indirect understanding of the system structure through its behavior. Figure 4.6 shows a stock and flow diagram for a

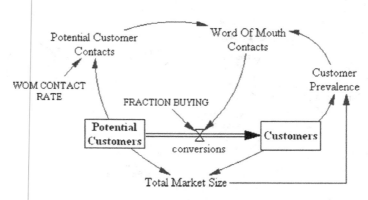

Fig. 4.6. **A stock and flow diagram for a generic marketing problem.**
Source: Vensim documentation, www.vensim.com.

commercial marketing problem. Here, the stocks (potential customers and customers) are connected by a flow link that is regulated by a rate (fraction of buying).

In this book we limit our analysis to causal loop diagrams, since a quantitative analysis of an engineering system is not within the scope of the research. Our aim is to foster insight and enable communication among stakeholders, and for that purpose a causal loop diagram is far more effective.

4.6.1 *Virtues of Causal Loop Diagrams*

Jim Hines, a professor of System Dynamics at MIT, expressed the following opinion on the virtues of causal loop diagrams (Hines, 2000):

> In consulting, I usually start with causal loop diagrams before going on to stock and flows. The exception is when I see immediately a very clear and important stock and flow structure (the iThink folks might call this a "main chain") in which case, I might dive into the stock and flow right from the start. In teaching the SD applications course here at MIT, we encourage students to start with causal loop diagrams. One reason for this is that students who start with stocks and flows often never complete any important feedback loops. Other reasons to start with causal loop diagrams include:
>
> - CLD's are usually more dramatic and hence capture the interest of students and clients alike (it's good to start with a bang).
> - Causal loop diagrams lead to insights on their own more frequently than stock and flow diagrams do. (Note, I am distinguishing between stock and flow diagrams and the simulation model.)

- Causal loops are easy to develop at a relatively high level of abstraction — this means that they can provide an overview of the system you are modeling, before getting down to the nitty gritty.
- Causal loop diagrams are fuzzier, so they can be drawn even if you are not yet clear on every single concept (this is a common state at the beginning of the project).
- Causal loop diagrams are cheap relative to simulation modeling (and cheap relative to an equation-level stock and flow diagram). This means you can more quickly get a comprehensive feel for the problem area. And inexpensively generate some initial insights.

4.6.2 *Drawbacks of Causal Loop Diagrams*

While causal loop diagrams have the advantage of simplicity, many systems experts warn against their misleading nature. The particular problems cited are the lack of distinction among types of links and the lack of characterization of the strength of links as well as their long-term polarity (Richardson, 1986). As illustrated in Chapter 2, the CLIOS process has tried to address these shortcomings by providing ways to distinguish links in terms of their characteristics, relative strength and stopping at individual link polarity assignments (thereby refraining from preassigning loop polarities). Yet we do realize that for the quantification of representations into quantitative models, a transition into a stock and flow structure or other model structures may be necessary. One could also argue that a system representation should only be used as a qualitative information and knowledge-organizing tool, rather than as the basis of a quantitative model. This is an interesting area for further research and lies beyond the scope of this book.

4.7   Experiments in Engineering Systems Representation

One of the hypotheses expressed in Section 4.5 was that the representation of an engineering system is far from objective, and depends on the values, cognitive limits and perspectives of those present during the representation. This hypothesis was explored in two experiments we did with MIT and Cambridge University students. The first experiment dealt with biases in identifying components and linkages in identical systems by different teams of graduate students, and the second dealt with the ability of experts to look beyond their specific parts of the system to encompass interconnections with other parts of the system outside the boundaries of their own analysis.

4.7.1   *Cognitive Limits of Identifying components and Linkages for a Unique System*

In Fall 2002, an assignment on the representation of the transportation air pollution system in Mexico City was handed out to graduate students in the "Introduction to Technology and Policy" course taught by Prof. Joseph Sussman at MIT. The assignment followed lectures on the basics of CLIOS representation and the Mexico City transportation-air pollution linkage.

Student teams had little or no prior knowledge of the system. All students were given written technical and institutional background material, and were asked to represent the transportation subsystem of Mexico City with CLIOS causal loop diagrams that would serve as a blueprint for modeling the transportation air pollution system. Preassigned student teams consisted of four to five individuals with similar educational backgrounds, most of whom were first-year graduate students in the Technology and Policy program at

MIT. Most had undergraduate degrees in engineering and 0–2 years of work experience. Half of the student group was American and the rest were international students. Women made up around 40% of the class, and were represented nearly equally in all teams. The time for the assignment was set at two weeks for all teams.

What these representations clearly showed was different emphases on which components were to be included, and what interconnections existed among them. If each of these transportation subsystem representations were used as the basis for a quantitative transportation air pollution model, they would result in different analyses and policy recommendations.

The focus of the representation of Group 1, shown in Figure 4.7, was the link between fleet size, fuel quality,

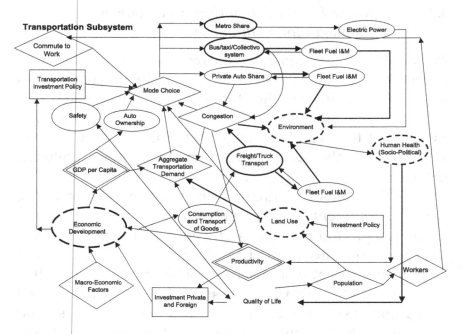

Fig. 4.7. **Group 1 representation of the Mexico City transportation subsystem.**

inspection and maintenance (I&M), congestion and their effect on air quality. Group 2 had its focus on the link between Metro electricity consumption, fleet age and fleet usage, congestion and freight transit (Figure 4.8). Group 3 took into account only the effect of congestion and Metro electricity consumption on the air quality (Figure 4.9). Group 4 added more sophistication by including policy levers that could affect air pollution through better emission standards and better urban planning (Figure 4.10). While most groups focused on the impact of congestion on the environment, three out of four explicitly addressed the loss of productivity as a main issue to consider when thinking about the air pollution problem.

In fact, if all of the above representations were integrated they would provide a better picture of the drivers and

Transportation Subsystem

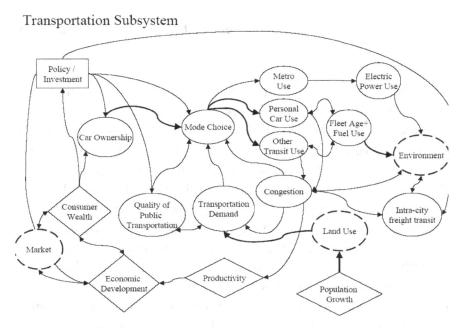

Fig. 4.8. Group 2 representation of the Mexico City transportation subsystem.

Transportation Subsystem

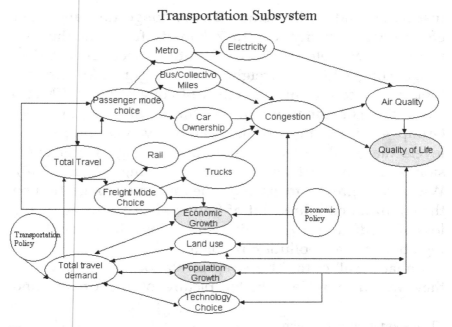

Fig. 4.9. Group 3 representation of the Mexico City transportation subsystem.

components that affect air quality in Mexico City. The combined representation also enables decision-makers to draw on the resources of stakeholders who would be more concerned with the economic impacts of congestion rather than its air pollution impact. This would enable dual-purpose strategies that could address both problems at the same time, with more institutional backing, than if the problem were framed only as an environmental problem. Table 4.1 shows some other quantitative differences among the representations.

What this experiment shows in general is that there is a cognitive bias that impacts both the number and type of components identified and the interconnections between them. The level of detail that is followed for particular linkages also varies within the individual representations as well as across

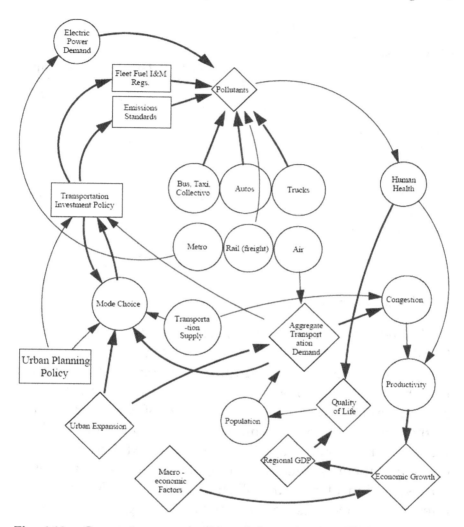

Fig. 4.10. Group 4 representation of the Mexico City transportation subsystem.

different groups. A counterargument can be made that as experts develop competency in their chosen field, they develop common mental maps which reduce these differences among different representations and the subsequent models.

Table 4.1.   Representation Characteristics

Characteristic/ Team	Group 1	Group 2	Group 3	Group 4
Number of policy levers	4	1	3	2
Common drivers	9	4	0	5
Other components	17	15	16	17
Total components identified	30	20	19	24
Total linkages identified	47	45	41	33
Link to component ratio	1.6	2.3	2.2	1.4

4.7.2   *Partial System Cognition*

A different experiment was performed with a group of Cambridge University students enrolled in the Technology and Policy program. The group of students included both British and international students, mostly with undergraduate degrees in engineering and 0–2 years of work experience. In the "Introduction to Technology and Policy" course (TP3) taught by Prof. Sussman, similar to that at MIT in the previous experiment, students were again asked to do a CLIOS representation of the air pollution problem in Mexico City after hearing lectures on the topic and being provided with written background materials on the subject. This time, each of the groups was assigned one "subsystem" to look at. In effect, each of the subsystems was made into a system for the groups to represent. These included preassigned systems of 1) freight transportation, 2) automobile transportation, 3) public transportation, surface (bus, taxi, colectivo), 4) public transportation, Metro, 5) PEMEX (the Mexican national energy company) and 6) electricity utilities.

Generally (with the exception of the group that was assigned PEMEX), the groups did well in identifying detailed components and linkages within their assigned systems. What was interesting however was the way they captured the interconnections of their assigned system with other parts that make up the Mexico City air pollution problem. Only two groups out of six captured most of the interconnections with other systems adequately. This illustrates the difficulty that field experts may face when there are multiple subsystem representations, with each looking only at one part of the system and not having the larger system to look at.

This experiment illustrated two main points:

(1) Choosing the system boundaries are crucial in addressing the right problem.
(2) It is difficult to capture all the necessary interactions between an area of focus in the system with others outside the system boundaries if not starting from a holistic system representation that addresses the problem as a whole.

4.8   Stakeholders, Conflict and Systems Representation

System representations are the foundations on which policy and strategy recommendations for large-scale engineering systems are designed. They reflect how we frame a problem, how we define the system boundary within which it is to be understood, how we identify key components and linkages, and how we measure its performance. It is based on our representations of the system that we identify alternatives for design or management of the system to improve its performance. As such, systems representation may be the most important part of engineering systems analysis and design. As the review of the literature and the above experiments suggest, stakeholder participation in the representation of an engineering system is imperative if we are to consider their concerns and knowledge within the decision-making process. Many conflicts in the process are the result of inadequate problem definition and performance measure identifications (i.e. what we consider good performance for the system?) in the representation. By involving stakeholders in the system representation process we can hope to build a solid foundation for the rest of the decision-making process.

Additionally, because of the biases inherent in the mental map of any one individual, one could argue that the more minds work on a representation, the more comprehensive and accurate it becomes. This is essentially the hypothesis of this book. But it would be myopic not to consider some of the challenges that have so far slowed the involvement of stakeholders in representation processes. In the next section, we will take a more detailed look at these challenges.

4.9   Challenges of Involving Stakeholders in Engineering Systems Representation

The old saying "Too many cooks in the kitchen spoil the broth" is not irrelevant in stakeholder processes. The incremental value of adding a new perspective is offset to some extent by *increased process complexity*. In this section we identify major challenges that have to be considered when designing a stakeholder-assisted representation process.

Increasing conflict: We said earlier that one of the major goals of involving stakeholders in the representation of complex engineering systems is to reduce conflict. Yet when we engage stakeholders in the representation process, we are *shifting* some of the conflict from the implementation and decision-making phase of projects to the design and analysis phase of engineering systems. Conflict in representation often takes the form of disagreements on what the "important" system components are and how they are interconnected. More importantly, the decision of which "performance metrics" should be included in the representation is subject to evaluative complexity. Here performance metrics refer to components that constitute the intermediate and final outputs of a representation, if it were quantified. In the next section, we will discuss where conflicts can arise in the representation process. However, it should be noted that in a well-designed collaborative representation process, the intensity of conflict is reduced substantially, and its destructive energy is channeled into creating a representation that is rigorous enough to better withstand criticism from different perspectives. Therefore, while we are adding to the complexity of the representation process, we are reducing the overall conflict level of the entire decision-making process. This overall reduction of conflict can often change the fate of the entire project from impasse and deadlock to seamless social and institutional implementation.

Process obstruction: There are always individuals within the stakeholder group who are unwilling or unable to work collaboratively with others. Many times they try to dominate discussions and create obstructions if not managed effectively. A good process has to ensure that its structure and ground rules are designed to minimize obstructionist efforts.

Heterogeneous knowledge and expertise backgrounds: When bringing together decision-makers, experts and citizen group representatives, the heterogeneity of stakeholder backgrounds is a given. Even experts from different fields could be considered laypersons when it comes to fields outside their expertise. One challenge is for the representation process to both be technically accurate, while being accessible to stakeholders at the same time.

Representation validation: Another important consideration is the validity of the representation created from a technical standpoint. This is primarily ensured within a well-designed process by having participating experts sign off on the jointly created representation. Yet it is still imperative to initiate an outside "peer review" process by independent experts who can vouch for the relative validity of the engineering systems representation.

Increased process time: One other issue with involving stakeholders is the perceived increase in the length of the representation process. While it may seem quicker to have a limited group of experts focus on the representation, they may have a harder time getting access to data outside their own reach, and to convince non-participating stakeholders of the merits of their representation. There is limited comparative empirical research on the relative lengths of stakeholder-assisted representation and expert-based representation processes that could shed light on how serious this drawback is. It can be argued that involving stakeholders from the beginning would reduce the overall time of the decision-making process (from problem definition to project implementation).

Increased representation and modeling costs: A more comprehensive representation is not essentially a better representation. Many experts make judgments on which parts of the system may be less important to consider. By reducing the size of the problem, they make its analysis more feasible. Many engineering systems experts are worried that the inclusion of stakeholder concerns in a representation may only add to the costs of the analysis without substantively improving the analysis. Furthermore, many of the social and economic considerations included by stakeholders may be difficult to quantify. In large-scale engineering projects, adding more details to the scope of the environmental impact assessment process can increase the costs of the technical process dramatically and prolong its completion. In order to address this concern, we should make a distinction between the representation as a whole and the part of a presentation that has to be quantified for decision-making. Often, many components in the system representation only serve as a context for decision-making, providing us with a perspective on the various impacts of an alternative. Much of the scope may also have a qualitative analysis component, which can be done in parallel by social science experts rather than technical experts. Additionally, many times stakeholders have funds at their disposal that is used for competing or adversarial expert analysis. Many government advisory agencies have access to data that would be an asset to the project. Other stakeholder groups may have budgets they could contribute to the analysis process, thereby reducing the financial burden of the additional technical analysis. This concept is known as "resource pooling", and is one of the advantages of the feeling of ownership that results from involving stakeholders in the decision-making process.

4.10 Designing an Effective Stakeholder-Assisted Representation Process

The stakeholder-assisted representation process includes a series of meetings in which a problem is defined, the system boundaries that can address it are selected, components and their interconnections are identified, and information on the problem and the relevant system are shared and organized within a system representation by the participants. Generally, for such a process to work well, formal and informal procedures and rules have to be defined in a way that facilitates the interactions of stakeholders in the representation process, and establishes clear guidelines on how information is shared and used in the representation. In addition to the initial design of the process, it is necessary for a facilitator to guide discussions, so that the process can actually progress. Any stakeholder process consists of a brainstorming and idea formation stage, which is divergent in nature. Since the goal of involving stakeholders is to agree on a common system representation, effective facilitation is needed to ensure that this goal is reached. This is particularly true for drawn-out lengthy processes and complex representations (Dean *et al.*, 1998).

There are essentially three distinct stakeholder-assisted representation approaches:

Indirect stakeholder involvement: Extracting inputs from individual stakeholders through surveys and interviews, as well as other means of extracting stakeholder inputs. Experts then construct a system model based on the inputs, which is then sent to individual stakeholders for feedback. This is the least participatory form of stakeholder involvement in representation. It will also be harder to involve stakeholders in the consequent quantification and alternative design stages of the decision-making process. On the other hand, it minimizes conflict in the representation stage, and shifts it to future stages of the decision-making process.

Direct stakeholder involvement: Stakeholders jointly create the system representation starting from scratch. This has the highest learning value for stakeholders, but is also the hardest to facilitate due to rapid convergence of dialogue and beginning of conflict. It may however create more trust among stakeholders after an initial period of time, reducing the conflict in the overall decision-making process.

Hybrid indirect-direct stakeholder involvement: This initially starts with indirect involvement, but once an initial representation is constructed, stakeholders are invited to refine it. One of the advantages of this approach is that with the initial representation available, stakeholder dialogue remains more focused than direct involvement. It is also more likely to reduce conflict at the representation stage, while allowing stakeholders to shape the final representation together. With the bulk of the representation ready from the beginning, the process will also become shorter, reducing the load on facilitation. The drawbacks may be reduced learning on the part of stakeholders, and reduced feeling of ownership for some stakeholders.

Selection of the appropriate approach depends on what trade-offs we are willing to make for the process. In cases where conflict is not as pronounced, the direct approach may be the most valuable. In cases where stakeholder inputs are desirable but no possibility of a stakeholder process exists, one could use the indirect approach. In this book, we have used the hybrid approach as a way to reduce the effects of obstructionism, reduce the load of facilitation and minimize conflict at the representation stage.

4.11   Limitations of System Representations as a Basis for Collaborative Processes

Despite the above-mentioned advantages of system representations serving as a basis for collaborative processes, there are serious limitations that a representation-centered process imposes on stakeholders. For one, there are many stakeholder concerns that cannot be expressed in terms of boxes and arrows. Feelings of injustice, of non-representation and moral wrong cannot be expressed adequately in a "logical" causal loop diagram. Additionally subjective issues such as aesthetics, fairness, etc. lose their real meaning when put into a causal loop diagram. Therefore, a system-representation-centered process does not allow for the more emotional concerns of stakeholders to be taken into consideration. This would limit the application of such a process in cases where the emotional load is far more important than the rational concerns.

On the other hand, we would argue that for some cases this framing of the problem, while imposing some structure on the discussion format, can help address many of the problems associated with collaborative processes, as seen from an engineering systems perspective. While there may be other venues and channels for stakeholders to express their emotional concerns, a system-representation-centered process could help uncover the underlying tangible interests of stakeholders within the technical process. To what extent this is feasible has to be seen in actual practice, but the basic approach of this research is that in addition to a change of attitude in experts, there needs to be a change of attitude among stakeholders as well for such processes to succeed. It is too early to predict whether such an approach is doomed from the start, or whether it can be useful in some cases but not in others.

4.12   Chapter Summary

In this chapter we looked at the concept of representation in general, and the concept of engineering systems representation in particular. We explored the role of values, beliefs and different types of cognitive biases in constructing system representations. We then looked at ways to involve stakeholders in the representation of engineering systems and identified challenges associated with such involvements. We introduced effective stakeholder involvement approaches for engineering systems representation, and discussed their relative merits. In the next chapter, we will look at the stakeholder-assisted modeling and policy design (SAM-PD) process that forms the core methodological contribution of this book. A major part of this process focuses on stakeholder-assisted representation of engineering systems.

Bibliography

Box, G.E.P. (1976) Science and statistics. *Journal of the American Statistical Association*, 71: 791–799.

Craik, K. (1943) *The Nature of Explanation*. Cambridge: Cambridge University Press.

Davidson, M.J., Welt, J., (1999) Mental models and usability Depaul University, Cognitive Psychology 404, November 15.

Dean, D.L., Lee, J.D., Pendergast, M.O., Hickey, A.M., and Nunamaker Jr. J.F, (1998) Enabling the effective involvement of multiple users: Methods and tools for collaborative software engineering. *Journal of Management Information Systems*, 14(3): 179–222.

Hines, J. (2000) The virtue of causal loop diagrams (CLD). Email message by Jim Hines, a professor at MIT, to the system dynamics listserv on February 23 (reproduced from http://world.std.com/~awolpert/gtr367.html).

Marks, D. (1999) Consciousness, mental imagery and action. *British Journal of Psychology* 90: 567–585.

Richardson, G.P. (1986) Problems with causal-loop diagrams. *System Dynamics Review*, 2(2): 158–170.

Rumi, M.J. (2004) *Tales from the Masnavi*. Trans. Mojaddedi, J. Oxford: Oxford University Press.

Chapter 5

Stakeholder-Assisted Modeling and Policy Design

The idea is to try to give all the information to help others to judge the value of your contribution; not just the information that leads to judgment in one particular direction or another.

— Richard P. Feynman

If the previous chapter could be considered to provide a conceptual framework for this book, this chapter can be taken as its core methodological contribution. After discussing the different aspects of stakeholder involvement in engineering systems decision-making at length from a variety of different angles, here we propose a process that aims to facilitate that involvement. The process, called stakeholder-assisted modeling and policy design (SAM-PD), combines insights from earlier discussions into a comprehensive process that facilitates stakeholder involvement from problem definition to post-implementation monitoring and adaptive management.

5.1 The Stakeholder-Assisted Modeling and Policy Design Process

As the literature indicates, ideally a "good" or desired outcome of a successful stakeholder process for engineering systems decision-making would have the following attributes:

- It would produce a package of holistic (system-wide) decisions based on the best available science (within an agreed timeframe and available resources) agreed upon by the overwhelming majority of key stakeholders.
- The policies would have taken into account the values and the local knowledge of the key stakeholders that have emerged in a collaborative process through facilitated dialogue.
- The policies would actually address the problem at hand effectively over time.
- The policies would be adaptive so as to integrate emerging scientific data and changes to the system and the environment.
- Implementation of policies would meet little resistance from the affected stakeholders and would ideally not result in extensive litigious action, resulting in robust solutions.

Based on these insights, we have designed the stakeholder-assisted modeling and policy design process. SAM-PD uses insights from systems thinking and alternative dispute resolution (ADR) to provide an integrated engineering systems decision-making process that enables stakeholders, decision-makers and technical experts to make collaborative decisions. The process is based on a holistic analysis of engineering systems within a collaborative framework. As noted earlier, SAM-PD uses a consensus-building process as its collaborative framework, and a CLIOS process for its systems analysis stage. Figure 5.1 shows the concept of the SAM-PD process as a synthesis of the two processes.

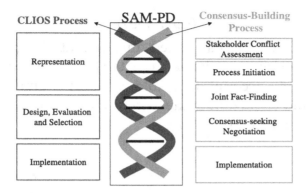

Fig. 5.1.   The SAM-PD double-helix, connecting the CLIOS process and the consensus-building process strands. The links holding the strands together include stakeholder-assisted representation and model-based negotiation, the participation level point (PLP) heuristic, the stake-power-knowledge (SPK) framework, discourse integration and pragmatic analysis.

SAM-PD in a Nutshell

In SAM-PD, representative stakeholders, decision-makers and technical experts jointly engage in defining the scope of engineering systems policy issues they have a direct interest in. They use a collaborative process to represent the system that addresses the issue at hand through a holistic systems analysis perspective that allows them to better understand the interactions among different parts of the system and between the different technical, social and economic layers of a system. Based on that representation, they collectively design alternatives and evaluate their effects on the system using the system model they created. Finally, in a consensus-seeking negotiation based on quantitative results from the model as well as

(Continued)

(Continued)

qualitative insights gained throughout the process, they forge policies that address the identified problem, taking into account stakeholder concerns and knowledge and cognizant of the uncertainties inherent in the scientific/technical analysis.

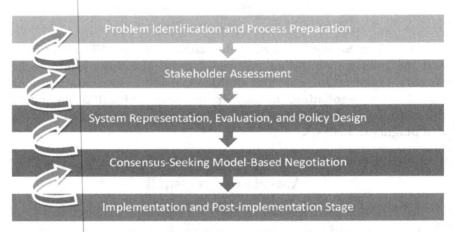

Fig. 5.2.   The SAM-PD process outline.

The PLP heuristic, the SPK framework, discourse integration, pragmatic analysis, stakeholder-assisted system representation and model-based negotiation will be discussed in detail later in this chapter.

5.2   Outline of the SAM-PD Process

SAM-PD is a five-stage iterative consensus-building process that uses a system representation as the basis of dialogue and negotiations among stakeholders. Figure 5.2 shows an outline of the five stages.

In the next sections, we will look at the detailed steps within each of the five stages.

5.3   Problem Identification and Process Preparation Stage

As seen in Figure 5.2, the first stage of the SAM-PD process is *problem identification and process preparation*. At this stage, a problem in an existing system is identified, or a new system is proposed. With the initial decision question defined, decision-makers assess the need for stakeholder participation and form a *convening* or managing group for the process. If a collaborative process is warranted, the convening group chooses a neutral who can prepare the groundwork for the actual process.

5.3.1   *Problem Identification/Project Initiation*

Engineering systems decision-making processes start when a new system is desired to serve a particular need, or when problems emerge in an existing system. Often, the official decision-making process starts after this point, with the identification of a problem in the management of an existing system or the official initiation of a new project.

Strategic resource management/regulation: Many existing large-scale engineering systems exhibit weak performance or exert unintended impacts on their environments at some point over their lifetimes. At such times, agencies and organizations managing important parts of the systems begin to think of interventions that would improve the performance of such systems, or address the problems that have emerged. For the type of engineering system we are concerned with in this book, that is, systems that have significant social and environmental impact, it is often a combination of public and private institutions that initiate the problem identification.

In these cases, government agencies with the mandate to manage and regulate different parts of the system decide to define/redefine resource management strategies after an actual set of problems or potential future problems are identified in the system, either by the public, media and/or the agency's experts. An example of this type of process is the strategic management of air quality in Mexico City, strategic management of spent nuclear fuel in the U.S., national and regional energy policy, and existing infrastructure improvements.

Permitting processes: Most newly proposed engineering systems need to go through some form of a permitting process. For large-scale projects, such processes often include an environmental impact assessment. In such cases, the developer or a government agency initiating the project has to go through a permitting process, where an assessment of the potential social, economic and environmental impacts of the project and available alternatives are studied before a permit for the project is issued. A permitting agency reviews the permit application and can approve or refuse an application for a new engineering system. In the U.S., this type of process takes the form of a National Environmental Policy Act (NEPA) process. A recent project undergoing a NEPA process is the proposed offshore wind farm project in Nantucket Sound, in Cape Cod, Massachusetts, that will serve as the central case study in this book.

In the SAM-PD process, we refer to organizations that have a regulatory or management role over the engineering system as decision-makers. It is important to realize that not all government agencies that manage parts of the system have a decision-making mandate. Many times, it is a single government agency that has the final say on the system in question. Decision-makers may have a stake in a specific outcome, but will be mostly concerned with the effective management of the system in a way that satisfies their organizational mandate. Normally, decision-makers follow established procedures for addressing a permit application, but there are less defined procedures for strategic resource allocation problems. Stakeholder participation in these procedures is often in the form of public hearings.

5.3.2   Stakeholder Participation Level Assessment

The PLP Heuristic: Virtually all complex, large-scale engineering systems have a multitude of stakeholders and would benefit from some level of stakeholder participation in the decision-making process. Decision-makers need to identify what level of stakeholder participation is necessary for the particular problem. As a heuristic tool, we have developed the participation level point (PLP) heuristic (shown in Table 5.1), which links system/stakeholder characteristics with the participation ladder categories proposed by Arnstein (1969). The premise of the PLP heuristic is that some problem/system characteristics increase the desired level of stakeholder participation.

The PLP heuristic provides a direction, not an answer. As such, it is always wiser to err on the side of higher stakeholder participation than to settle for lower stakeholder participation levels. If the PLP of a system is 4 or higher, stakeholder participation in the system representation and modeling stage is advised. Given that the different questions in Table 5.1 do not necessarily carry the same weight in different contexts, decision-makers need to use their own judgment to assess whether this heuristic is appropriate for their particular system. Paradoxically, there may be a need for a basic, more limited stakeholder consultation to determine whether stakeholder involvement is necessary and if so at what level. It is important to note that in any

Table 5.1.   The PLP Heuristic

Step 1: Examine system characteristics	Yes	No
Is the system in question spread over multiple jurisdictions?	1	0
Does the problem affect a multitude of heterogeneous stakeholder groups?	1	0
Has the issue already stirred visible controversy?	1	0
Are cost distribution issues important?	1	0
Is all the funding necessary for building/ managing the system available to the decision-makers/project developers?	0	1
Is uncertainty in scientific information a source of controversy?	1	0
Are environmental justice issues relevant?	1	0
Is there distrust of the decision-makers' ability to adequately represent stakeholder interests?	1	0
If marginalized in the decision-making process, do stakeholders have the ability to adversely impact the implementation or management of the project/system in a significant way?	1–2	0
Do some stakeholders have access to useful information/data or financial/human resources they would be likely to share if they were involved?	1	0
Is adaptive management of the system over time imperative?	1	0
Is significant process obstruction by stakeholders likely if they are involved?	−1 or −2	0

Participation level points (PLP)	**Sum**

Step 2: Determine level of participation (based on a modified Arnstein's ladder (1969)	**PLP**
Public participation in final decision	9 or above
Public participation in assessing risks and recommending solutions	6–8
Public participation in defining interests, actors and determining agenda	4–5

(Continued)

Table 5.1.   (*Continued*)

Restricted participation (feedback in public hearings, commenting opportunities)	3
Public right to object	2
Informing the public	1
Public right to know	0

decision-making process there are different levels of participation for different stakeholders, depending on their stake, power and knowledge. The PLP only points to the highest level of participation desirable in the project. As part of future work on SAM-PD, we plan a calibration of the PLP heuristic with past engineering systems cases, to make it a more reliable heuristic.

If the PLP of a system is lower than 4, then there is little use for the SAM-PD process in the decision-making process. SAM-PD is suitable for systems where the degree of conflict, uncertainty, distrust and information heterogeneity has evolved to an extent that makes more aggressive stakeholder participation a necessity. Sometimes, a system's PLP changes over time, with more controversy and uncertainty emerging over time. If such changes in the system characteristics are expected, the PLP can be evaluated with the potential changes in mind.

5.3.3 *Choice of Convener*

If the PLP heuristic indicates the need for a collaborative process (PLP $\geq$ 4), the decision-makers need to identify a suitable convener group. The convener group is the entity that manages the collaborative process by formally inviting stakeholders and bringing them together, providing facilities for the duration of the stakeholder process, providing funding for major parts of the process and choosing the neutral. For these reasons, the convener group should have credibility, authority and trustworthiness in the eyes of the potential stakeholders. It should also possess sufficient funds and resources to ensure that the process can be carried out through to the end.

5.3.4 *Choice of Neutral*

The *neutral* is the main person in charge of stakeholder identification and selection, stakeholder conflict and value assessment. The convener group chooses the neutral to perform a conflict assessment for the project. The neutral then assembles a team and starts the stakeholder identification process.

It is wise to choose the neutral from outside the convening organization, preferably a professional in the field of negotiation and conflict resolution, with a robust knowledge of stakeholder conflict assessment practices. Given that it is desirable to preserve knowledge in the process, and that stakeholders will interact with the neutral during the conflict assessment process, it would be advantageous for the neutral to be a prime candidate for the facilitator position later in the process, but this doesn't always have to be the case.

5.4   Stakeholder Assessment Stage

The stakeholder assessment stage is one of the most important stages in the SAM-PD process. The success or failure of the process may depend on which stakeholders are engaged and at what level.

5.4.1 *Stakeholder Identification*

Engineering systems often impact a multitude of stakeholders, some obvious, some less so. The obvious ones are the people advocating a project/management strategies, the vocal people or groups who oppose that proposition as well as the government agencies that have the mandate of making decisions on the issue. Usually, however, there are a number of other stakeholders who are likely to be affected by and therefore concerned about any decision that is made, and may try to reverse the decision or block its implementation, if their concerns are not integrated into the decision-making process (CRC, 1998).

Given the limitations on how many stakeholders can physically participate in a collaborative process, it is necessary for the neutral and the convening group to assess at what levels individual stakeholders or their representatives should be involved.

Effective stakeholder identification is therefore imperative to determine who will be directly or indirectly affected, positively or negatively, by a project or a system management plan, and who can contribute to or hinder its success. It is important for the project sponsor/system manager to be comprehensive in identifying and prioritizing all relevant stakeholders, including those who are not usually present at the table (Susskind and Thomas-Larmer, 1999). Those identified will then need to be consulted to varying degrees, depending on their potential impact on the system, as well as their potential to contribute to the policy process through knowledge, resources or compliance with implementation. Stakeholders can be categorized based on their influence/ power, stake and knowledge:

- *Decision-makers* (high stake, high power and differing levels of knowledge): Representatives from organizations that have a mandate to manage some part of the system or issue a permit for a new project, as well as other organizations with mandates over other systems interconnected with the target system, whose help is required in effectively managing the system.

- *Stakeholders with economic/political influence* (high stake, medium to high power and differing levels of knowledge): These include affected industry, private corporations, landowners, labor unions, nationally recognized and highly organized NGOs and other groups with strong political influence.
- *Knowledge-producers* (low stake, low power and high knowledge): Technical experts, engineers and consultants working in academia, technical consulting firms, local, state and federal science agencies and scientific and technical offices of government agencies and scientific arms of NGOs that have a stake in the process, but have no specific mandate.
- *Other affected stakeholders* (high stake, low power and differing levels of knowledge): These include smaller groups of stakeholders directly or indirectly affected by system management strategies or the proposed project. These can include less organized neighborhood groups, local environmental groups, small business owners, etc., depending on the type of system or project that is initiated.

The SPK framework (Figure 5.3) provides a rough mental guideline for the stakeholder classification process. Stakeholders can be assessed by their stake, power and knowledge (expert or local) on the decision. Stakeholders with high stakes in the collaborative process, even if they lack any power or knowledge, can add legitimacy and community acceptance. Stakeholders with high knowledge can add to the scientific/technical/contextual validity of the analysis, while stakeholders with power (that is, mandate or resources) can increase the viability of the process. Stakeholders with lower stake, power and knowledge can be involved through feedback systems, information websites, media releases and outreach campaigns.

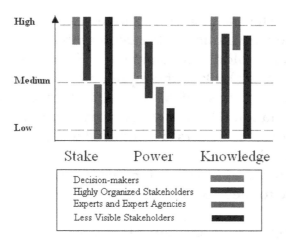

Fig. 5.3.   The SPK framework.

Of course it is important to realize that such a categorization, while useful as a rough map, should not be the exclusive criteria for selecting stakeholders for participation, given that even smaller actors can sometimes be effective in undermining a process.

5.4.2   *Stakeholder Value Assessment*

Once a basic stakeholder list is prepared, it is imperative to establish the stakeholders' interests/values regarding the system/project, eliciting their views on the system/project, and the issues they would like to have considered in any policy process. This stage will help generate a set of information on the basis of which a tentative system representation can be built.

As mentioned in Chapter 4, in this book we use a *hybrid* direct-indirect stakeholder involvement process. We indicated that one of the advantages of this approach is that with the initial representation available, stakeholder dialogue remains more focused than direct involvement. It is also more likely to reduce conflict at the representation stage, while allowing stakeholders to shape the final representation together. With the bulk of the representation ready from the beginning, the process will also become shorter, reducing the load on facilitation. The drawbacks may be reduced learning on the part of stakeholders and reduced feeling of ownership for some stakeholders.

The following approaches for eliciting stakeholder inputs are available to the neutral:

- stakeholder conflict assessment surveys,
- interviews,
- media articles and press releases,
- stakeholder websites, and
- formal and informal hearing transcripts.

Stakeholder value assessment survey: Once a basic stakeholder list is prepared, it is imperative to establish the stakeholders' interests/values regarding the system/project, eliciting their views on the system/project, and the issues they would like to have considered in any policy process.

Media articles, press releases and interviews: In many cases it is difficult to reach all the key stakeholders for commenting. There are some indirect ways of considering the views of stakeholders on the system. One of the ways to expand the range of stakeholder inputs is to study newspaper articles, television programs and press releases that exist on the system in question. In addition to positions, there are often statements that express the underlying values and concerns of stakeholders that can be extracted.

Stakeholder group websites: Most organized stakeholder groups in the U.S. and other developed countries have some of their views presented on their websites. These are usually far more comprehensive than those found in newspaper articles although like newspaper articles, websites only represent the voice of those already vocal. However, in many cases it can be a good supplementary source of information, should it not be possible to access some stakeholders for direct input elicitation.

Many stakeholder organizations also have comment sections on their websites where individual stakeholders (usually those who support the position of that particular stakeholder group) leave feedback or comments.

Formal and informal hearing transcripts: In many cases, formal or informal hearings are held at different stages of decision-making. Transcripts of these hearings, when available, can shed further light on stakeholder views on the system. Given that such hearings are usually open to the public, they are a good way of capturing stakeholder inputs from less organized stakeholder groups.

5.4.3 *Selection of Process Participants and Individual Participation Levels*

The answers to the questions in the previous step, along with the initial categorization of stakeholders, should provide a basis for the selection of participants for the collaborative process. Stakeholders not included in the initial interviews but mentioned by a considerable number of other stakeholders should be contacted and interviewed. Stakeholders in each category should be ranked according to their importance to the process and chosen based on the criteria of authority, political power, intensity of interest, potential for knowledge contribution, potential for resource provision and potential to undermine agreements if excluded. This is essentially a case-by-case decision, but given the structure of the collaborative process, the process would be most effective if the number of participants did not go beyond a certain limit. While there is no fixed limit on the size, having larger groups can result in unmanageable group dynamics, while very small groups can result in many of the different stakes not being covered by those present. There should be a balance among the four categories of stakeholders in the core group present at all stages in the process.

While it is imperative to have the most crucial stakeholders participate in the process as a core group, some of the sessions could be held with additional stakeholders who can contribute in particular stages, but not be present at all stages. Some may be chosen to participate in all stages of the process, while others may be asked to provide feedback in different stages of the process and be kept informed. Every effort should be made to have the most crucial stakeholders in the process from the very beginning, but if at any time a key stakeholder is identified who has been left out, they should be consulted and if possible included in the process. The process should be designed so that inputs from stakeholders not directly participating in the process can be considered for inclusion at any time.

After the selection occurs, selected stakeholders are invited to participate in the collaborative process. Many of the selected stakeholders will be skeptical whether or not to participate in the process, unsure of how it might benefit them. Here, it is the task of the neutral to present a compelling case for the benefits of the collaborative process. Selected stakeholders should be invited to attend the introductory session, where the decision whether or not to proceed with a collaborative process is made. Given that they still have the option not to participate after the introductory session, many selected stakeholders may agree to attend such a session. Before the introductory session, the neutral provides the selected stakeholders with a list of all the participants and a summary of the interviews, so individual participants can understand the interests, concerns and positions of other participants, categorized under each set of questions.

5.4.4 *Choice of Facilitator*

In the first face-to-face stakeholder meeting of the SAM-PD process, selected stakeholders who have agreed to participate come together for an introductory session aimed at building initial trust and getting to know other stakeholders and their interests and points of view. The convener presents some background material on the basics of the consensus-building process, and explains what the group can expect as an outcome of such a process. The group of stakeholders jointly decides whether or not to proceed with the process. Individual stakeholders may opt out of the process. If the remaining participants choose to proceed with the willing group of participants, the group can then proceed to choose a neutral facilitator (who can be the neutral chosen previously by the convener or any other person agreed on by the group). The facilitator is the person responsible for facilitating dialogue amongst stakeholders in all subsequent stages of the collaborative process. The ideal facilitator for such processes is a person who is competent in negotiations and conflict resolution theory and practice, and has a basic understanding of the system/project in question, and is known by stakeholders to be objective and neutral to the outcome. As the term "neutral" applies, the facilitator should have a record of professional facilitation and clear impartiality in the eyes of the various stakeholders. Once chosen, the facilitator initiates the next stage of the collaborative process, which is the joint fact-finding stage. Once the facilitator is chosen, the ground rules for the process have to be set. These include how sessions will be conducted, how decisions get made and how communication between sessions is established. It may also be useful to establish a neutral information repository for the entire group to deposit information about the system, as well as proposals for strategies and alternatives (Susskind and Cruikshank, 1987).

In a consensus-based process, the usual decision-making rule is by consensus of all those present. Given that this may result in one party sabotaging the process, it would be useful to agree to overwhelming majority. In the end, while some parties may not agree with individual decisions, a consensus is sought on the package of decisions produced by the whole group. In other words, consensus is actively sought and encouraged, but it is not the prerequisite for reaching final agreements.

5.5   Extracting Contextual Knowledge from Stakeholder Statements

With the direct and indirect stakeholder input solicitation methods discussed in Section 5.4 it is possible to extract the components necessary to build a CLIOS diagram of the different physical subsystems, and the institutional sphere. In this section we propose using insights from *discourse integration* and *pragmatic analysis*, approaches that are used to understand the meaning of written statements within the field of linguistics, as a way to extract representation-related information from stakeholder comments.

5.5.1   *Discourse Integration and Pragmatic Analysis*

It is not surprising to say that language is an incomplete means of communication. This is more the case when we do not have additional cues, such as body language, tone and interaction to understand an individual's statements. This is particularly important for cases where stakeholder inputs are elicited from surveys, newspaper articles or other written material, where stakeholders cannot be asked to clarify their remarks.

In order to reduce the subjectivity of the process of converting stakeholder statements into system representation elements, we can use insights from two basic approaches in linguistics called discourse integration and pragmatic analysis. The use of insights from these approaches is based on "Principles of Critical Discourse Analysis" by Van Dijk (1993), which is considered one the classic works dealing with understanding citizen discourse and underlying values.[13]

Pragmatic analysis focuses on the structure of an individual sentence. Information is extracted by looking at the position of words, and the relationships within the sentences themselves. Discourse integration reinterprets the sentence with the context of the statement in mind, that is, who has expressed the sentence, what other sentences came before it or within what context the statement was expressed. The entities involved in the sentence must either have been introduced explicitly or be related to entities that were, and the overall discourse must be coherent. This is mostly the case for both surveys and written transcripts of hearings and websites. A combination of these two will allow us to keep only those parts of the statements that contain useful information for the system representation.

Generally stakeholder statements can contain an intricate combination of implicit and explicit values and positions, information on the system and "mental maps". A mental map is the way a person perceives the outside world and the way its components interact and function. Our purpose is to extract the underlying values, the proposed drivers of those values and qualitative and quantitative information that can

---

[13] Van Dijk, T. (1993) "Principles of critical discourse analysis", *Discourse and Society*, 4(2): 249–283.

be used to represent the system and quantify it (model) later in the process.

5.5.2 *Applying Discourse Integration and Pragmatic Analysis to SAM-PD*

Let's consider the following statement:

> Supporters of the project say that any bird deaths would be minimal compared with the millions of birds that die each year colliding with skyscrapers and cellphone towers. Opponents, meanwhile, have said any bird deaths are unacceptable Beth Daley, "Report on possible risks from wind farm fuels ire", *The Boston Globe*, October 17, 2004.

We use pragmatic analysis to look at the structure of each of the sentences.

> Supporters of the project say that any bird deaths would be minimal compared with the millions of birds that die each year colliding with skyscrapers and cellphone towers.

Here, the structure of the sentence is as follows:

"X (supporters of the project = subject, stakeholder groups) *say* (plural subjects) *that* (objective expression) (qualifier = any) (value/performance measure = bird deaths) (adjective = minimal) (conditional = if compared to) (value/performance measure = birds that die each year) (by means of action/ component = colliding with skyscrapers and cellphone towers).

> Opponents, meanwhile, have said any bird deaths are unacceptable.

"Y (opponents = subject) *meanwhile* (relationship = have said) (qualifier = any) (value/performance measure = bird deaths) *are* (position = unacceptable).

We can extract from the excerpt that there are two differing views on the particular topic.

From *discourse integration* we can connect the contexts of the two sentences, inferring positions from the proponents also. The implicit position of the proponents is that bird deaths would be minimal and thus, based on a presupposition of affirmation, acceptable. We can also derive the causal components of the "bird deaths". In the first sentence we see skyscrapers and cellphone towers explicitly mentioned, but unless we pay attention to the subject "supporters of the project", we cannot derive the implicit causation that comes from the project. The general knowledge that "the project" refers to a wind farm is part of the discourse integration that comes from previous parts of the article.

In summary, the following information in Table 5.2 can be extracted from the statement using discourse integration and pragmatic analysis.

Table 5.2.   Extracting Contextual Information from Stakeholder Statements

Type of Information	Information
Performance measures	Bird deaths per year (total versus wind-farm induced)
Positions	Values: only zero is acceptable (opponents); has to be measured relative to other projects (proponents)
Causal component	The wind farm (through its turbines) (opponents and proponents); skyscrapers and cellphone towers (proponents)
Data	Bird deaths from other man-made structures in the millions

Not every statement contains useful information for systems modeling purposes, but nearly every statement provides information on the context within which discourse integration and pragmatic analysis can be performed. It is important to realize however that there is the possibility that stakeholders may not talk about the underlying values openly. Therefore, the emphasis on the totality of statements, rather than emphasis on individual statements, is important in understanding stakeholder interests and concerns. This approach can also be dangerous, since it has the underlying assumption that rational concerns that can be assessed objectively are always at the root of stakeholder positions. This assumption of rationality may not hold most of the time, but is necessary when coupling stakeholder inputs to a more or less objective systems analysis. Also, many stakeholders may use fake concerns to mask their real concerns. For example, if stakeholders feel that aesthetics may not be considered as important by the decision-makers as much as environmental impact, they may express concern over environmental impact while the only thing they care about is aesthetics.

5.5.3 *Converting the Contextual Information into a System Representation*

Using the CLIOS notation, cellphone towers and skyscrapers are external drivers (beyond the boundaries of the current system), the wind farm is a policy level (we can decide whether or not to allow it) and bird deaths (both total and as a result of the wind farm) are performance measures. The acceptable thresholds can then be discussed during the performance metric refinement with stakeholders. Figure 5.4 shows the representation of this statement.

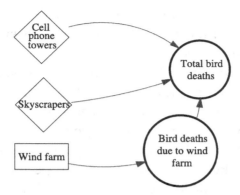

Fig. 5.4.   Causal loop representation of the stakeholder statement.

5.6   System Representation, Evaluation and Policy Design Stage

The system representation, evaluation and policy design stage of the SAM-PD process essentially consists of system and goal definition, system component and linkage representation and characterization, and design, evaluation and selection of alternatives. As such it corresponds to steps 1–9 of the CLIOS process. In this section, we will look at the details of this stage while mapping the SAM-PD process onto the CLIOS process.

5.6.1   *Problem Refinement/System Definition (CLIOS Step 1)*

In this step, problems with the system have to be clearly described and a summary of issues and system goals prepared. The first step of the joint fact-finding process is to determine the scope of the problem to be studied. The scope determines where the system boundaries are and what issues/areas need to be addressed. The boundaries can be both in terms of geographical area covered, as well as the components of the system that are considered. In the engineering systems literature, the scoping of a system is often called a system representation.

Traditional environmental impact assessments mainly focus on risk assessment in defining the scope, but it is possible for the stakeholders to take into account benefits resulting from alternatives as part of the scope of the analysis as well. A good example is the case of the proposed offshore wind farm in Nantucket Sound, where many potential benefits of clean energy for Massachusetts and the United States could be used alongside the potential risks in the scope of the scientific analysis.

It is also possible to take into consideration non-risk-related issues such as aesthetics arguments and social effects of projects, where a pure scientific analysis may not be necessary, but where options to address these concerns and reduce their potential impact on stakeholders would require expert knowledge. As an example, the impact of erecting wind turbines in Nantucket Sound on tourism in the region or real estate prices could be assessed to some extent using similar projects in other regions that have been allowed. Whether or not such considerations are taken into consideration in the scope of the problem is up to the group of stakeholders, but if the actual concerns of NIMBY (not in my backyard) advocates is not addressed in some form, it may be expressed in terms of emphasis on uncertainty in the science, which may lengthen the scientific analysis process, or make it more difficult to reach an agreement in the negotiations phase.

The important thing to note is that the scope will be heavily affected by who is present at the table in the collaborative process. While decision-makers are required by law to define a minimal scope for the problem, technical experts have to make sure that the scope is sufficient or possible to evaluate, while other stakeholders will try to address their own concerns in the scope (NRC, 1996). Usually, different stakeholders highlight the parts of the system that are directly of interest to them, or those which if analyzed would favor their positions. This is essentially a value-based judgment, and can result in conflict. The challenge for the facilitator is then to reframe or redefine the issues in terms of interests, which are usually negotiable, rather than positions, values or needs, which usually are not. This is called "interest-based" framing, and is an approach proposed by Fisher *et al.* (1991). They argue that by focusing on the interests rather than on positions, there is a higher possibility of a robust agreement, since it may be possible to find a solution, which satisfies both parties' interests. Once the underlying interests are identified, they will be discussed in the group. The opposing sides will be more motivated to take those interests into account if they feel their interests are also being taken into consideration. The aim of discussions is to find possible solutions that satisfy the interests of as many stakeholders as possible (Fisher *et al.*, 1991).

The problem identification and summary of goals should capture the concerns and needs of policy-makers, managers and stakeholders and how the system is or is not responding to them. The initial problem definition by the convener group is thus refined to reflect stakeholder perspectives on the issue.

Much of this information will be available through the input elicitation activities in the stakeholder conflict assessment stage.

The stakeholder survey should include the following questions for individual stakeholder organizations, to serve as the basis for the system representation, evaluation and policy design stage of the SAM-PD process:

- What is the stakeholder's view of the system boundary/ the scope of the project?

- Which part of the system/project are they interested in?
- What is their organizational interest/mandate regarding the system and how does it impact their position on the project/system management strategies? Does their organization favor a predefined position, or a predefined set of strategies? If so, how does that position serve their interest?
- How does the system affect them at present and how do they think it will impact them in the future? (stakeholders with influence, other stakeholders)
- What are the most important issues they see with the project/system? What do they think could be done to address these issues?
- What are the institutional relations that govern the system? (decision-makers)
- What information do they possess on the project/system? What information do they believe is necessary, but missing? What capacity do they have for further information gathering? What are the resources they have at their disposal to contribute to the management of the system/ evaluation of the project?
- What is the approximate timeline in which the decision has to be made? Is the timeline flexible or fixed? Can the decision be staged? (decision-makers)
- How would they want to participate in the decision-making process? Would they like to be present at all stages, or be kept informed of all the stages, or would they like to provide feedback once the recommendations are opened up for public comments? What do they think of a joint fact-finding process as an alternative for the decision-making process?
- What is the internal decision-making mechanism for the organization? Who has the authority to negotiate in a potential joint fact-finding process?

- Who are the other stakeholders that should participate in the decision-making process? Also, who if not involved could undermine the quality, legitimacy or outcome of the joint fact-finding process?

For each project/system, questions specific to the system should be substituted whenever appropriate. The neutral then synthesizes the interviews into a value assessment report that can be used by the different stakeholders to understand the scope of values, interests and knowledge that other stakeholders hold.

The answers to the questions in the previous step, along with the initial categorization of stakeholders should provide a basis for the selection of participants for the collaborative process.

5.6.2 *Initial System Representation (CLIOS Steps 2–5)*

Before stakeholders come together for the collaborative system representation, we propose that an initial system representation is constructed based on stakeholder inputs. In our view, this provides many advantages over starting the system representation from scratch with the stakeholders present. We referred to these advantages in Chapter 5, but for the sake of convenience, we will briefly review them here:

(1) Providing stakeholders with a sense of common accomplishment right from the start.
(2) Providing stakeholders with an initial holistic perspective of the problem.
(3) Focusing the discussion from early on, and facilitating convergence.
(4) Effective utilization of stakeholder times and avoiding early frustration.

It often happens that participants overlook issues which are important to others but are not important to them. Particularly, when stakes are high, the number of issues that people think are part of the problem in dispute tends to increase. However, if the most important issues are not identified, it will be impossible to develop successful solutions to the conflict (CRC, 1998).

There are basically two main issues that have to be addressed at the representation level:

(1) Have all the important issues been identified?
(2) Should every issue that any stakeholder considers as important be part of the analysis?

In addressing the first question, it is important to have a diverse enough set of stakeholders at the table, such that a comprehensive coverage of issues is explored. The role of technical experts at this stage is crucial, since many of the issues that have to be considered in the analysis may be salient and not so obvious. On the other hand, having all the issues that are discussed in the analysis may make it impossible to analyze the system in time or at all, which is a point that the second question raises. While it is imperative to be as inclusive as possible, there is no easy way out of this. Essentially, the inclusion of issues should result from an overall agreement by the group that the issues are important enough to be considered. The facilitator's role in making sure that the group considers each issue carefully is essential in keeping the process from alienating those stakeholders whose proposed issues may not be taken into consideration as part of the scientific analysis. An important consideration is how to capture the relationships between the different issues in the system, and how the links among the different components can be represented.

Figure 5.5 is an outline of the steps leading from stakeholder input elicitation to system representation.

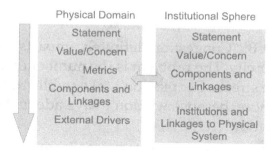

Fig. 5.5. Converting stakeholder inputs into system representations.

We generally start with the values/concerns (performance measures) and work our way back to external drivers step by step. In this way we are constructing a representation that has outputs that are important for stakeholders to consider in the decision-making process.

The basic methodology used for the system representation is the one outlined in the CLIOS process, as described in Chapter 2.

In order to illustrate how this is done, we consider the transportation/air pollution example below.

Illustrative Example: Transportation/ Air Pollution System

Step (1) Extracting contextual knowledge from stakeholders

Statement A: "We are really concerned with the congestion in the city, we believe that it is adversely impacting the productivity of the city. I believe the problem is that we don't have sufficient highway infrastructure. The government should build more highways".

(Continued)

(Continued)

Statement B: "Congestion definitely impacts air pollution in our city. The problem is the number of cars that are on the streets. People don't take public transportation anymore, because we don't have adequate public transportation".

Step (2) Extracting values/concerns from statements

Statement A: Value/performance measure: Productive time loss

Stated Driver: Congestion

Stated secondary driver: Infrastructure availability

Statement B: Stated value/performance measure: Air pollution (health + smog)

Stated Driver: Congestion

Stated secondary driver: Public transportation availability

Step (3) Determining system performance metrics (Table 5.3)

Table 5.3.   System Performance Metrics

Values/Drivers	Metric
Air pollution	CO, CO_2, NOx, NMHC emissions (tonnes/year), PM10 concentration, ozone concentration (ppm) Premature deaths due to pollution (people/year)
Productive time loss	Average hour/person/trip, total hours per day, $/hour/trip, $/day
Congestion	Average vehicle speed (km/hr)
Infrastructure availability	Km of arterial roads

(Continued)

(Continued)

Step (4) Building the systems representation diagram (Figure 5.6)

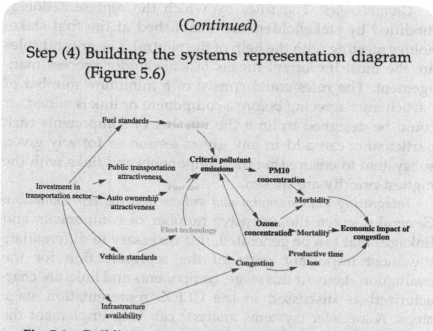

Fig. 5.6. Building a systems representation diagram based on stakeholder statements.

At the initial representation stage, we assume that all stakeholder concerns should be included in the system representation. It is up to the stakeholders to negotiate which parts of the system can or should be quantified (modeled) later on when they review the representation.

5.6.3 *Stakeholder-Assisted Representation*

Once the initial representation is created, participants chosen by the neutral for the collaborative process can review and refine it, by adding new components and links or agree to remove unnecessary components and links from the representation.

Ground rules: The rules by which the representation is modified by stakeholders are established at the first stakeholder meeting with the help of the neutral. The ground rules are the most important means of stakeholder process management. The rules could consist of a minimum number of participants agreeing before a component or link is added, or could be designed to limit the number of components each participants can add in any given session or for any given subsystem to ensure that the components and links with the highest priority are chosen.

Integrating pbibliography and values into the representation: Generally, given the extensive number of components and linkages that can be generated, it is necessary to differentiate the most important parts of the representation for the evaluation stage. At this stage, components and links are characterized as discussed in the CLIOS representation stage steps. A modeler (systems analyst) can help implement the changes in the representation software, that can be projected at all times on a screen visible to participants.

Identifying interactions between actors and physical system: Once a commonly agreed representation emerges, it is time to look at which organizations and institutions have control, influence and knowledge on different parts of the system. This is an important part of the preparation for the working group formation. Here the concepts of Class 2 links or "Projections" in the CLIOS process can be applied (see Chapter 2).

5.6.4 *Evaluation/Quantification of the Representation*

An important part of the evaluation step is to identify what information is required to assess the system both in its current state, as well as the potential impacts of the new project or proposed management strategies.

At this stage the following questions have to be answered by the group using the cumulative knowledge of the participants: What data is needed to describe the current status of the system? How much of this information is available? If some data exist, are they sufficient to make an informed assessment of the decision? If not, what type of data is required to make such an assessment? What are the uncertainty levels and what levels of uncertainty are acceptable for such an assessment? How can further baseline data be acquired, and what resources are available within the group to obtain additional information that will contribute to an assessment of the system that is acceptable by the stakeholders? (Rebori, 2000).

The presence of representative stakeholders, decision-makers and technical experts can provide a clear picture of who in the group has the information, expertise or resources that could be used to assess the system. This is one of the strengths of the joint fact-finding process compared to traditional processes, where funding and information is limited to that of the decision-making entity.

Strategy/Alternative generation: The generation of alternatives is usually done in brainstorming processes carried out by the facilitator. Brainstorming is a collaborative technique for generating new and innovative ideas and solutions to a problem. A facilitator initiates brainstorming by asking the parties to suggest ideas for solving the problems identified in the scoping process. Judgment about the merits of the proposed solutions/strategies is withheld until later. The facilitator usually lists the ideas in a way that is visible to the participants, helping them keep track of what has been said, and to build on earlier suggestions. This often results in creative solutions to problems that no one person or one side would have been likely to develop on their own (CRC, 1998).

Once all the ideas are out there, the facilitator helps the group to narrow down the alternatives to a manageable size, by grouping related proposals together and making the alternatives consistent in terms of level of detail and underlying assumptions. The participants develop and implement a joint strategy for answering the key policy questions, based on jointly agreed methodologies. Here, participants do not have to reach agreement on the methodologies for every issue. Their primary goal is to clearly separate the issues upon which they can agree from those which are still subject to debate. Points of mutual agreement can then help as a basis for continuing the analysis of disputed issues (CRC, 1998). As in all other stages, the role of the facilitator is crucial to the success of the process. This process may take several sessions, depending on the difficulty of reaching an agreement on issues.

Working group formation and fact-finding: Once the alternatives to be explored are agreed on by the group, working groups need to be established to explore the system, both in terms of the baseline status and in terms of analyzing the potential impact of strategies or alternatives on the system. While for smaller problems two to three working groups may be sufficient, in a complex technical system more working groups may be required. Working groups should be sufficiently diverse to incorporate different stakeholder interests, so that no working group is considered representative of a particular stakeholder view. While it is impossible to explore all the issues with the full group, it is important that the full group be still engaged in some form in the joint fact-finding process.

Questions that will shape the working group structure by which to assess impacts are: What additional data will be needed to assess the impact of the new project or proposed strategies? If new data are necessary, who will collect the data? Who pays for data collection? How will outside experts be selected? What methodologies should be applied? Who will manage the data gathering and coordination? What kind of information repository will be necessary? Will the collected information become public knowledge or kept confidential until agreements are reached? What are the timeframes for collection, analysis and reporting? And finally, who will own the data once collected? (Rebori, 2000)

Basically, a fact-finding committee is formed from expert members of each of the conflict parties, in addition to independent experts hired by the stakeholder group and decision-makers who have some knowledge about the issues being explored. Experts may have to be paid through the convener or through the pooled resources provided by the stakeholder group. The working groups have the task of working out the important facts and eliciting relevant knowledge from the literature, as well as other commonly agreed sources. This kind of collaborative approach will result in a level of interaction that would not likely occur under other circumstances where each of the experts represented their respective interests, thus shifting away from "adversarial science" (Schultz, 2000). If there is expertise missing in the ranks of the participants for a particular area, the group can invite outside experts that it commonly agrees on to form working groups on that issue.

With the shared information sources and expertise, factual knowledge on the system, in the form of technical literature, agency findings and consultant expertise, is very likely available to some extent for the system in question. The problem then becomes one of agreeing that the information is relevant for decision-making. An important challenge is that in high-stake, highly uncertain science-intensive conflicts, the facts in dispute are focused on the most uncertain areas.

There may be general agreement on the majority of the facts, but the process can stop when it comes to facts in dispute that are difficult to quantify or establish within an acceptable uncertainty range. That alone may result in a stalling of the process. Therefore, it is imperative to have the working groups focus on producing a new document synthesized from the available literature that reflects not just where consensus was achieved but also where disagreement on factual issues remains or where there is irreducible uncertainty. This kind of bottom-up approach, in addition to giving the group a definite goal, allows the group to focus their efforts on the facts instead of on positions, allowing for invention or consideration of new solutions in the process (Schultz, 2000).

Communicating results to the group: In terms of communication of results, it is important that the working groups report to the full group on a regular basis, so that the group as a whole can provide feedback on whether or not to proceed with further studies and exploration of issues. Effective factual communication is important, since this allows non-experts to offer possibly fresh insights, forcing experts to examine a set of problems in a new way. Additionally, this can lead to better understanding of the system as a whole by the stakeholders, which may further help them understand other points of view on the system. (Schultz, 2000).

After several intervals of fact-finding and reporting, once the results of the different working groups are deemed sufficient for decision-making purposes, the facilitator and the working groups draft a combined document summarizing and synthesizing the findings. The final document is then presented to the stakeholders for evaluation of the different alternatives. Often, the document is accompanied by a variety of scientific and technical computer models that can assess the impact of different strategies/alternatives on the system.

Transparency of process to the public at large: It is imperative for all the information and transcripts from the meetings to be available to the public. This can be done by establishing a stakeholder process website, where progress in every session is reflected in an accessible format to those not directly participating. The website should also allow the public to provide comments and feedback on the process. The neutral should provide participants with copies of feedback given by the public at the beginning of each session, to ensure that participants can use the feedback in refining the discussions. The balance of where to draw the line is delicate and needs to be explored, as integrating the feedback can slow down the process immensely.

Level of quantification/evaluation: It can be argued that there exists a level of quantification which, accompanied by a qualitative analysis and caution about uncertainties, can provide a better understanding of the strengths of linkages to the stakeholder and is therefore an essential step of the analysis. On the one hand John Sterman argues for the importance of quantitative models for a group understanding of a problem at hand. While a qualitative model can potentially increase a group's information processing capacity, he points out that it is dangerous to draw conclusions on the dynamics of a system solely based on diagrams, a position which can hardly be refuted, given the wide range of evidence (Sterman, 2000). On the other hand, advocates of the use of qualitative modeling have argued that in a number of cases quantification may either decrease the model's relevance for an audience or can even be dangerously misleading as well.

However it is clear that selecting between different strategies requires some form of quantification, given that there are always trade-offs involved. Policies are then designed jointly with stakeholders based on the qualitative and quantitative insights gained from the model they helped create. In addition to technical and economic feasibility, the social feasibility of options is brainstormed with stakeholders present. Using the jointly developed model, negotiations can then be carried out to reach agreement on a set of policies.

5.7 Consensus-seeking Negotiation

5.7.1 Facilitated Negotiation on Alternatives and Consensus-based (or Overwhelming Majority) Agreement

Once the working groups have submitted the final report, and before actual evaluation by the full group takes place, it is important to agree on objective criteria by which to analyze alternatives. Objective criteria refer to factors that are used to evaluate a decision or possible outcome. Objective criteria will help move the group from the joint fact-finding mode to a decision-making mode. People usually support objective criteria during a collaborative process because criteria are not tied to specific positions (Rebori, 2000). However, the objective criteria would probably be based on the important performance metrics that the stakeholders have specified in the previous steps. Objective criteria can be categorized into three broad categories (Godschalk, 1994):

Technical criteria: To test different strategies/alternatives, the group can establish criteria such as levels and coverage of service, performance standards, resource requirements, or degree of impact of the project. As an example, in the case of a new offshore wind power project, the criteria can be the amount of electricity produced, the amount of greenhouse gas emissions prevented, the cost of electricity produced, the number of potential bird fatalities, the number and severity of navigational problems, the number of fish affected, whale population changes in the area, etc.

Social and community criteria: These can measure the societal and economic impacts of the strategies/project. In the case of offshore wind power, this could be net employment change, change in real estate prices, change in tourism revenue, change in fishery income, etc.

Value-based criteria: This is the trickiest set of criteria as they are hardest to quantify. They can incorporate some NIMBY values, as well as other sociopolitical values that are hard to quantify. While it may be difficult to compare different strategies/alternatives on the basis of these criteria directly, it is important to take them into consideration, and find ways to capture them in terms of negotiable items. In the case of offshore wind power, these could include the number of visible wind turbines on the horizon, the height of the turbines as seen from the shore, the number of lights that can be seen from the shore (all an effort to capture the aesthetics value of an unfettered ocean view). Of course, in many NIMBY arguments, the issue is mostly binary in nature, but it may be possible to address the concerns of some of the groups by addressing some of these issues in terms of objective criteria.

As indicated in the previous section, it is useful to have models, where different assumptions for different packages of alternatives can be evaluated easily, without having to redo the entire joint fact-finding process. Many of the alternatives may be generated after the stakeholders have a better picture of the system as a whole, a result of the final fact-finding document coming together. Unlike other processes, technical experts and experts stay on with the full group to assess the impact of the different alternatives/strategies that the group comes up with. For this purpose it is useful to have an integrative model of the entire system, which can predict the impact of one change on different components of the system simultaneously, thus capturing some of the complex dynamics that a system may exhibit. Using an overall systems framework, and an adequate system representation it is possible to integrate the knowledge created into one coherent model. While some of the alternatives/strategies may not be quantifiable, it is useful to see how they would impact the system qualitatively (Mostashari and Sussman, 2003).

Using the joint fact-finding document and the models, the group then proceeds to look at the most promising alternatives identified in previous steps under the range of uncertainties and examines the costs and benefits from different stakeholder perspectives while also exploring any barriers to implementation. It does so based on the objective criteria that were agreed on by the participants. Here, however, the main focus of the discussions will be on the uncertainties about the system, which the working groups were not able to reduce to generally acceptable levels. There are two ways to proceed in this stage. One is to devise experiments that may provide more certainty and more knowledge on the issue. The other is to proceed with the given uncertainty range and negotiate *contingent agreements*, which specify specific actions that would be taken to ameliorate the potential consequences of the problem to risk levels acceptable to the group as a whole. The former will probably serve as a good delaying tactic for those interested in stalling the project/strategies, as it will be costly and time consuming to do in most cases. The latter is normally undesirable for decision-makers and in case of permit processes would pose additional risk for the developer.

After the stakeholders have explored different strategies in the strategy analysis stage, sets of policies can be designed jointly with stakeholders based on the qualitative and quantitative insights gained from the model they helped create. In addition to technical and economic feasibility, the social feasibility of options is explored with stakeholders present.

Negotiations using the developed model are carried out to reach agreement on a set of policies. Stakeholders can brainstorm on implementation procedures for policies in agreement, focusing on how costs and benefits would be distributed and how responsibilities would be distributed.

As indicated previously, the goal of the consensus-building process is not to have agreements on every single issue, but to agree as a group on a package of strategies/alternatives that are acceptable as a whole. As in all other negotiations, the success or failure of reaching an agreement depends on the individual participants' best alternative to a negotiated agreement or BATNA (Fisher *et al.*, 1991). It is up to the facilitator and the evolving group dynamics to shape the BATNA in a way that an agreement seems desirable. Generally, if the process can be sustained for a long time, the time and resource investments, as well as the change in relationships due to collaboration, should help in the final process, creating a momentum for reaching a consensus-based agreement.

The list of alternatives is then narrowed down to one package of solutions which are fine-tuned until all the parties at the table agree on them. A helpful strategy is for each participant to propose several possible packages that are acceptable to them. Once all the packages are proposed, the group can work together to develop several variations in an attempt to develop a mutually preferred alternative. In this way, an agreement can be packaged without anyone having to make imbalanced concessions. Given the difference in priorities, it should be possible to find ways to accommodate most participants' interest such that they would be able to reach an agreement (ODRC, 2000).

The concept of contingent agreements can make it more attractive for some stakeholders to agree on a package. But for such an agreement to succeed, the participants have to develop a basic level of trust. Additionally, contingent agreements have to be documented with great care to ensure that they are not misused by any side, given that they are often not part of the conventional agreement documents. At the end, if a consensus is not reached, an overwhelming majority can also be sufficient for the agreement as a last resort.

Even if the participants themselves can reach an agreement, there is the additional issue of the stakeholders at large. It is imperative to communicate the analysis and the decisions back to the main body of stakeholders frequently during the process, meaning those who were not directly involved but had some form of representation in the process. The participants, as the representative stakeholders, have to ensure that their constituents understand the reasoning behind the decisions and have access to the analysis performed by the group. This is a difficult task, given that non-participants have not developed the same level of understanding or trust necessary to understand why this is the best possible agreement they can get. If any one of the groups represented in the consensus-building process disagrees at this stage, they will likely refuse to sign the agreement, and the agreement may well fall apart (CRC, 1998). Clearly, the skills of the facilitator can be the key to success. If sufficient alternatives/solutions are generated in the previous steps, the facilitator has a more open hand in highlighting areas of possible agreement.

Using a "single text" method to draft the written agreement is a useful way of getting closure. In this method, the group works on the agreement by moving through a single document together, with the facilitator either assigning preparation of the text to an expert who is not a stakeholder, or through having a small group of stakeholders prepare a draft on behalf of the entire group. The draft would have no legal status until the group reviews and refines the draft to reach agreement on a single final text. Agreement reached on any section of the document is taken as tentative until a final agreement on the entire document is reached. This can help by preventing individual stakeholders from bringing in alternative drafts of agreements, which can derail the process of reaching an agreement (ODRC, 2000).

If an agreement is reached, the group has to decide on an implementation schedule and resource allocations and divide responsibilities among the different participants.

5.8 Process Effectiveness and Validity Assessment Through Peer Review

Stakeholder-assisted processes are probably more costly and may take a longer time than traditional engineering systems analysis processes. It is therefore important to evaluate whether they have been more effective in addressing the problems, and whether the representations, model and recommendations that have resulted from such processes are considered valid from an outside perspective, and whether they can be implemented.

5.8.1 *Perceived Effectiveness of Process*

Surveys can be used to identify whether a particular stakeholder-assisted modeling and policy design process has been effective in the minds of decision-makers, experts and other stakeholders. Each survey can be tailored to its particular audience (decision-makers, advisory agencies, citizen groups, experts, private sector) within the participant group. The following are examples of questions that could be asked in each survey addressed to different stakeholder categories (Table 5.4). The 1–5 effectiveness rating scale allows stakeholders to distinguish between their perceived

Table 5.4.   Sample Questionnaire for Perceived Process Effectiveness Assessment

Criteria "In your view…."	Effectiveness of Process (1–5) 1 = Not at all effective 5 = Very effective	Importance of Criteria 1 = Not at all important 5 = Very important
1. Did the process help the decision-makers fulfill their mandate or their organizational duties effectively?		
2. Did the process reduce the cumulative potential conflict of the decision-making process effectively?		
3. Did the process involve stakeholders in decision-making effectively?		
4. Did the final recommendations address institutional and technical issues effectively?		
5. Did the involvement of stakeholders inform decision-making effectively?		
6. Did the involvement of stakeholders increase resources (funds, expertise and information sources) for decision-making?		
7. Did the involvement of stakeholders help communicate the rationale of the decision-making process to the public at large effectively?		
8. Did the involvement of stakeholders help the transparency of the decision-making process and increase public trust in the decision-makers effectively?		

(Continued)

Table 5.4. (*Continued*)

9. Did the final recommendations reflect the public interest effectively?
10. And, most importantly, are the final recommendations likely to address the problem at hand effectively?

effectiveness of each criterion. The importance of the criterion in the eyes of the participant filling out the survey allows for proper weighting of the criteria in assessing subjective effectiveness.

We emphasize that the answers to these questionnaires will be subjective, and can only measure the perceptions of the individuals answering them. But on a cumulative basis and over a large number of participants, the weighted scores can reflect the overall perception of the effectiveness of the process, while allowing for distinction of perceived effectiveness in the eyes of particular stakeholder groups.

Expert acceptance, in particular, of the validity of the recommendations is essential for the credibility of the process to non-participant stakeholders and the community at large. A process that is inclusive but relies on inadequate technical grounds could be perceived to be as undesirable as an exclusive process with good science/technical analysis. Even if all the participating experts agree on the validity of the process and its recommendations, it is important to have external feedback on its validity as well.

5.8.2 *External Validity of Process and Recommendations Through Peer Review*

After the final recommendations have been drafted, it is useful to have non-participant experts, decision-makers and the public at large review the process and its recommendations before it is officially published as the final agreements. Therefore, similar to a commenting period for an environmental impact assessment report, one can present the findings and recommendations at different public hearings, through websites and through active requests for peer review by experts. For that to happen, accessible documentation on the process should be available. There are two types of validity we are concerned with:

Process validity:

- Was the process valid within the context of the laws, regulations and mandates?
- Was the process sufficiently inclusive?
- Were the decision rules acceptable and adequate?
- Were the points of disagreement and the opinions of the dissenting participants adequately included in the final report?
- Were the expert working groups formed effectively?
- Was the process sufficiently transparent?
- Was the public given sufficient means of contributing to the process while it was in progress?

Validity of expert analysis and recommendations:

- Was the system representation used for the process accurate and valid?
- Were the methodologies used to evaluate the system representation valid?
- Were the assumptions made for the expert analysis valid?
- Were the different uncertainties adequately addressed?

Once feedback is received the group reconvenes to decide whether there are grounds to rework parts of the process, or if the objections could be addressed without major changes. The final opinion of a diverse set of stakeholders, experts and decision-makers outside the participating group can then be integrated into the final report that is published.

5.9   Implementation and Post-implementation Stage (CLIOS Steps 10–12)

"If policy adoption is courtship, implementation is marriage".[14] All the stakeholder-assisted analysis and strategy design in the world is in vain unless it addresses the actual system issues realistically and facilitates the implementation of the engineering systems project.

5.9.1   *Implementation Schedule, Monitoring and Enforcement Design*

Once the basic elements of an agreement are agreed on, the design of the implementation phase and monitoring is in order. There is often so much emphasis on reaching an agreement that the implementation phase receives too little attention, a fact that can erase all of the achievements of the collaborative process. Here the implementation schedule, resource commitments by the individual stakeholders within the specified timeframe, and optionally contingent clauses need to be refined and spelled out in a written document. The parties should consider agreements on all issues as binding. Sometimes there is a need for institutional change if the strategies are to be implemented. These changes are more difficult to achieve than strategies for the physical system.

---

[14] Weimer and Vining (1999) Introduction to *Policy Analysis: Concepts and Practice*, 3rd edition (New Jersey: Prentice Hall).

Implementation strategies should reflect an understanding of the realities of both the technological and organizational complexity, the scale of the project, the limited time for implementation and the boundaries of agency mandates and influence over member agencies. Additionally, they should reflect organizational and technological strategies to deal with potential system failure and maintenance after deployment. If the system performance fails to improve, or if serious problems emerge for a new system that were not anticipated, one should return to the system representation to assess which parts of the system behaved differently from expected an which components and linkages were identified or evaluated incorrectly. One could then devise strategies to address the problem from a new perspective. Unfortunately, real life is not so simple since time and money are normally in short supply for revisiting problems.

There may be many reasons for failure to implement a stakeholder-assisted engineering system strategy, even one that is well designed to address existing concerns. Some of these include:

- funding uncertainty,
- political uncertainty,
- lack of flexibility in strategy,
- lack of robustness of strategy,
- failure of established measures in new contexts,
- unpredictable human interactions with technology,
- unpredictable environmental impacts of technology.

A good implementation strategy is one that addresses all of these potential factors in one way or another. Since it is impossible to predict all the possible ways an engineering system can behave in the long term, it would be wise for stakeholders to accept responsibility for an *adaptive manage-ment* of the engineering system post-implementation.

5.9.2   *Adaptive Management*

Adaptive management can be defined as

> The iterative process of designing and implementing man-agement activities in a manner that allows the scientific basis for management plans to be rigorously tested. The primary objective of adaptive management is to develop a better understanding of the systems being managed and to apply that knowledge in a way that allows the manager to continue to learn and develop better management practices.[15]

While it is extensively used in the context of environmen-tal systems, it is a concept that is crucial to the management of any large-scale engineering system with emergent behavior.

C.S. Holling first used the term "adaptive management" in his book *Adaptive Environmental Assessment and Management* in 1978. Adaptive management is based on adaptive process control theory, emphasizing learning control devices. Essentially, adaptive management improves its strategies based on gradual feedback over time, fine-tuning the process in a way that increases experience-based learning (McLain and Lee, 1996). For complex systems with emergent behavior, adaptive management is the only way to ensure the sustain-ability of a system over long periods of time.

The SAM-PD process enables adaptive management of engineering systems by creating a system model that can

---

[15] Wildlife Crossing Engineering and Biological Glossary, http://www.wildlifecrossings.info/glossary.htm.

later be refined when the behavior of the system does not conform to the existing model of the system. In this way, the system model is refined over time, helping not only the management of that system, but adding to our knowledge of how similar systems should be managed. The existence of a system model enables stakeholder groups to go back and improve their understanding of the system behavior and propose new solutions that address the emergent problems based on newly gained insight. For instance, a system model of an offshore wind energy facility in Nantucket Sound could serve as the basis of a system representation for offshore wind energy facilities anywhere in the world. Any new insights into the system representation can then be quickly implemented in the system model, enabling other similar systems managing groups to act quickly to avoid similar problems. This is the power of a *jointly* created diagrammatic representation of a system, and counts as one of the most important advantages of the SAM-PD process.

5.10   Chapter Summary

In this chapter we proposed the idea of a stakeholder-assisted modeling and policy design process, and described its approach to involving stakeholders in engineering systems analysis, design and management. We looked at how the five-stage SAM-PD process would map onto the CLIOS process proposed by Dodder *et al.*, and explored its different steps parallel to that process.

Most of the ideas in this chapter were theoretical and relatively abstract. In the next chapters we will explore SAM-PD in action, when we look at the actual case of the Cape Wind Offshore Energy Project, where actual stakeholders were called on to participate in a SAM-PD process. With the Cape Wind project still ongoing while this book is being written, the case study concludes at the end of the systems representation stage.

Bibliography

Arnstein, S.R. **(1969)** A ladder of citizen participation. *American Institute of Planners Journal*, 35: 216–224.

CRC **(1998)** Consensus building. International Online Training Program on Intractable Conflict, Conflict Research Consortium, University of Colorado, USA.

Fisher, R., Ury, W. and Patton, B. **(1991)** *Getting to Yes: Negotiating Agreement Without Giving In.* New York: Penguin.

Godschalk, D. **(1994)** *Pulling Together: A Planning and Development Consensus Building Manual.* Washington, DC: Urban Land Institute.

Holling, C.S. (ed.) **(1978)** *Adaptive Environmental Assessment and Management.* New York: John Wiley and Sons.

McLain, R., and Lee, R.G. **(1996)** Adaptive management: Promises and pitfalls. *Environmental Management*, 20(4): 437–448.

Mostashari, A., Sussman, J. **(2003)** Stakeholder-assisted modeling and policy design process. International Conference of Public Participation and Information Technology, Cambridge, MA, November.

ODRC **(2000)** *Collaborative Approaches: A Handbook for Public Policy Decision-Making and Conflict Resolution.* Oregon Dispute Resolution Commission.

Rebori, J. **(2000)** Joint fact finding. Managing Natural Resource Disputes, No. 8, University of Nevada Cooperative Extension Publications, 05/2000.

Schultz, N. **(2000)** *Consensus Building and Joint Fact-Finding.* Boulder, CO: Conflict Research Consortium, University of Colorado.

Sterman, J. **(2000)** *Business Dynamics: Systems Thinking and Modeling for a Complex World.* Boston: McGraw-Hill/Irwin.

Susskind, L., and Cruikshank, J. (1987) *Breaking the Impasse: Consensual Approaches to Resolving Public Disputes.* New York: Basic Books.

Susskind, L.E., and Thomas-Larmer, J. (1999) Conducting a conflict assessment. *In Consensus Building Handbook, Sage Publications; (August 1999).*

Van Dijk, T. (1993) Principles of critical discourse analysis. *Discourse and Society*, 4(2): 249–283.

Weimer, D., and Vining, A.R. (1999) Policy Analysis: Concepts and Practice, 3rd edition. New Jersey: Prentice Hall.

Chapter 6

The Cape Wind Offshore Wind Energy Project

You cannot NIMBY anywhere, any time, and expect to have electricity everywhere, all the time.

— Norris McDonald[16]

The ocean and bays that surround us are perhaps our town's most important and defining natural resource and it is these unspoiled waters that are the very essence of Cape Cod. We are a community of people drawn to the sea as sightseers, swimmers, sailors, fishermen or beach-combers. We are thankful for, and jealously seek to protect, the open space of the ocean around us. There is no other part of our community that offers more sweeping vistas, wildlife diversity and a place of refuge from the steady march of development.

— Barnstable Land Trust

The Cape Wind project, a proposal to build the first offshore wind energy facility in the United States, has become one of the most controversial large-scale engineering projects in recent American history. The controversy over the uncertain

---

[16] Comment at Cape Wind DEIS hearing, December 2004 in Cambridge, MA. Norris McDonald is a founder and president of the African American Environmentalist Association.

long-term environmental, economic and social impacts of the project has polarized the Cape and Islands region of Massachusetts, and has captured regional and national headlines. The case is a good example of a complex, large-scale, integrated, open system, with a technological system interacting with a social system and an ecosystem under large amounts of uncertainty. As such, it will serve as an illustrative case study for this book. In this chapter, we will present an introduction to the case, its main issues and its historical background.

6.1 Project Timeline

In November 2001, Cape Wind Associates proposed to build a 420 MW offshore wind farm in a 24-square-mile area on Horseshoe Shoal in Nantucket Sound. The project is to be located beyond the three-mile limit of state waters in federal waters on the outer continental shelf (OCS). Under Section 10 of the Rivers and Harbors Act, the U.S. Army Corps of Engineers is the federal agency mandated to regulate all structures and work in navigable waters of the United States. The first application filed by Cape Wind Associates in November 2001 called for the installation of a single scientific data tower. That was approved by the Corps in August 2002. Upon approval of the data tower permit, project opponents sued the U.S. Army Corps of Engineers in federal court over its jurisdiction on the project. The federal court threw out the suit in September 2003. The opponents appealed the decision. The Corps determined in 2002 that an environmental impact statement (EIS) was necessary according to the National Environmental Protection Act (NEPA). Scoping meetings were held in 2002, where the public was given the chance to comment on the scope of the EIS. The draft environmental impact statement (DEIS) was released by the Corps in November 2004. The DEIS was challenged by the opponents of the project, as well as by the U.S. Environmental Protection Agency. Three public hearings were held in various locations, again giving the public the opportunity to comment on the content of the DEIS. In March 2004, the Corps was in the process of compiling public comments and releasing the final environmental impact statement. The initial permitting agency timeline had planned for the decision to be made by July 2003. The delay is an indication of the degree of controversy that has surrounded the project. Figure 6.1 shows the initial timeline proposed by the U.S. Army Corps of Engineers. The actual approval by the federal government was given in April 2010.

6.2   Legal Context for Offshore Wind Energy Development in Massachusetts

At the federal level, there have been efforts at encouraging renewable energy production. Among these, in the Energy Policy Act of 1992, a production tax credit (PTC) was made

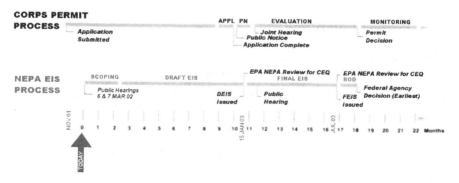

Fig. 6.1.   Initial timeline of the Cape Wind permitting process.
Source: U.S. Army Corps of Engineers.

available to entities that engaged in new renewable energy production through solar, biomass, wood chip, geothermal and wind electric power production. Additionally, at the state level, there are currently two laws encouraging renewable energy production. These are the Massachusetts Renewable Portfolio Standard (RPS) and the Massachusetts Renewable Energy Trust Fund (RETF) that are intended to promote the renewable energy portfolio of the state. The RPS required in 2003 that 1% of energy provided to consumers come from renewable energy sources. This will increase annually by 0.5%, requiring Massachusetts to produce at least 5% of its energy from renewable sources by the end of 2010 (Watson and Courtney 2004).

However, with regards to the usage of federal or state waters for wind energy facilities there are no clear laws, giving rise to controversy on how to regulate private use of public resources. While the Minerals Management Service (MMS) as part of the U.S. Department of the Interior has been in charge of assigning leases to offshore oil and gas facilities, they have currently no mandate to manage such leases for offshore wind energy facilities such as the Cape Wind proposal.

6.3   Project Overview

The proposed wind farm would have 130 turbines, down from an initially proposed 170 turbines. Each wind power generating structure would generate up to 3.2 MW of electricity and would stand up to 260 feet above the water surface. The power would be transmitted to shore via a submarine cable system consisting of two 115 kV lines to a landfall site in Yarmouth, Massachusetts. The submarine cable system interconnects with an underground overland cable system, where it will interconnect with an existing

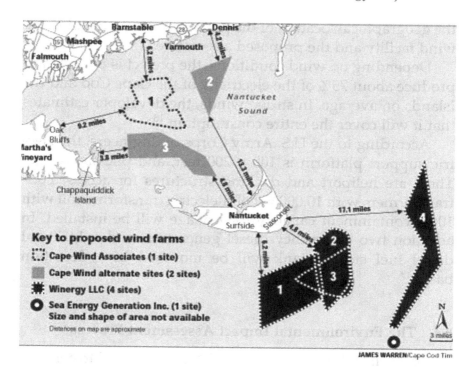

Fig. 6.2. Geographical location of the proposed Cape Wind project off Cape Cod, Massachusetts, (and alternatives).

Source: James Warren, Cape Cod Times, http://www.capecodtimes.com (last accessed October 19, 2004).

NSTAR 115 kV electric transmission line for distribution.[17] With the exception of two transmission cables and a portion of a proposed "wind wake buffer zone", the project will be located beyond the three-mile limit of state waters in federal waters on the outer continental shelf (OCS). Figure 6.2 shows

---

[17] Department of the Army, Corps of Engineers, "Intent to Prepare a Draft Environmental Impact Statement (DEIS) for Proposed Cape Wind Energy Project, Nantucket Sound and Yarmouth, Massachusetts, Application for Corps Section 10/404 Individual Permit", November 2001.

the geographical location of the proposed site for the offshore wind facility and the proposed alternatives.

Depending on wind conditions, the project is projected to produce about 75% of the electricity of the Cape Cod and the Islands on average. In strong winds, the developer estimates that it will cover the entire consumption.[18]

According to the U.S. Army Corps of Engineers, the electric support platform is 100 × 200 feet, and pile-supported. There are heliport and docking structures for access. Four transformers with 10,000 gal of dielectric transformer oil with 100% containment capacity for leakage will be installed. In addition two emergency diesel generators, and a 1,000 gal diesel fuel storage tank will be mounted in a detention basin.[19]

6.4   The Environmental Impact Assessment Process

Based on the NEPA process, Cape Wind was asked to compile an environmental impact statement as part of its application. The U.S. Army Corps of Engineers determined the scope of the EIS in conjunction with Cape Wind, inputs from federal and state agencies as well as public scoping comment periods. Tables 6.1–6.3 show the details of the scope of the report that was requested by the Corps.

Alternative sites: The U.S. Army Corps of Engineers has listed five alternative sites (the proposed site included) that will be assessed in the environmental impact statement. These are presented in Table 6.4.

---

[18] Cape Wind Associates, www.capewind.org.
[19] Adams, K. (Permit Manager), "Massachusetts Technology Collaborative Public Meeting Briefing", January 8, 2005, U.S. Army Corps of Engineers.

Table 6.1. Scope of the EIS: Impact on Avian and Marine Habitat, and Impact on Fisheries

Category	Issues to be Studied	Methodology to be Used
Avian habitat	— Current use of the final alternative sites by birds as baseline data — Species, number, type of use, and spatial and temporal patterns of use — Issues to be addressed include (1) bird migration, (2) bird flight during storms, foul weather, and/or fog conditions, (3) food availability, (4) predation and (5) benthic habitat and benthic food sources — Information derived from other studies, providing a three-year baseline data set — Endangered species impact on piping plover and roseate tern	— Existing, published and unpublished research results, especially research that describes long-term patterns in use — New field studies undertaken for this EIR/EIS. Data on use throughout the year, especially through November for migratory species, and under a range of conditions. Methods: Remote sensing through radar and direct observations through aerial reconnaissance and boat-based surveys — Data gathered through radar to be validated with direct observations — Known impacts to birds from former or current wind turbine generators (WTGs) and other tall, lighted structures (such as communications towers)

(Continued)

Table 6.1.   (*Continued*)

Category	Issues to be Studied	Methodology to be Used
Marine habitat	— Vibration, sound, shading, wave disturbance, alterations to currents and circulation, water quality, scouring, sediment transport, shoreline erosion(landfall) and structural habitat alteration — Northern right whale, humpback whale, fin whale, harbor seal and grey seal, loggerhead sea turtle, Kemp's Ridley sea turtle and leatherback sea turtle	— Assessment of (1) species type, life stage, and abundance; based upon existing, publicly available information; (2) potential changes to habitat types and sizes; and (3) the potential for turbines as fish aggregating structures. The study should assess potential indirect impacts on fish, mammals, and turtles that may result from changes in water movement, sediment transport, and shoreline erosion
Fisheries	— Assessment of potential impacts on specific fishing techniques and gear types used by commercial and recreational fishermen — Multiple-use conflict The potential for indirect impacts such as changes in fishing techniques, gear type and patterns will need to be included	— Review of existing literature and databases to identify and evaluate commercial and recreational fish data and abundance data in Nantucket Sound — Data to be reviewed should include National Marine Fisheries Service (NMFS) Commercial Data, NMFS Recreational Data, Massachusetts Division of Marine Fisheries Commercial Data, Massachusetts Division of Marine Fisheries Trawl Survey Data and supplemented with intercept surveys

Source: U.S. Army Corps of Engineers EIS Scope Document.

Table 6.2.   Scope of the EIS: Impact on Other Ecosystems and Physical Systems

Category	Issues to be Studied	Methodology to be Used
Benthic	— Sufficient information to compare alternative marine sites and to provide a general characterization of the benthic habitat of the final sites. (*Data on the Benthic Macroinvertebrate Community*)	— Assessment and additional data collection as described in the Benthic Sampling and Analysis Protocol (April 18, 2002)
Interactions between benthos, marine and avian food cycles	— Interconnections between the benthic, fisheries and avian resources — Predator-prey interaction data	— Noise and vibration impacts on fish and mammal habitats and migration — Assessment of the magnitude and frequency of underwater noise and vibrations, and the potential for adversely affecting fish and mammal habitats and migration — Assessment of fish and mammal tolerance to noise and vibrations, with particular emphasis on noise and vibration thresholds that may exist for each of the species

(Continued)

Table 6.2.   (*Continued*)

Category	Issues to be Studied	Methodology to be Used
Aviation	— Lighting requirements, radar interference and radio frequency interference — Lighting scheme will need to minimize impacts to birds while also providing for safe aviation	— FAA analysis
Communication	— Possible impacts to telecommunication systems — microwave transmission — Impact on installation of the wind turbine generators between Martha's Vineyard, Nantucket, and the mainland on existing transmission paths — Impact on boater communications devices	N/A
Navigation	— Commercial and recreational navigation impacts need to be addressed specifically for construction, operation and maintenance and decommissioning. — Cable installation activities to be included — National security issues	— U.S. Coast Guard risk analysis

(*Continued*)

Table 6.2. (*Continued*)

Category	Issues to be Studied	Methodology to be Used
Socioeconomic	— Impacts on electricity rates and reliability in New England — Explanation of any public funding and any applicable tax credits — Impact on local economy — Environmental justice issues — Educational and tourism impact	
Electric and magnetic fields (EMF)	— Data on potential human health impacts of exposure to 60 Hz EMF and potential impact of EMF produced from wind turbine generators and their associated cables	— Identify populations that could be exposed to 60 Hz EMF greater than 85 mG, including humans, fish, marine mammals and benthic organisms
Air and water pollution	— Compliance with the requirements of the Clean Air Act for construction and operation phases — Potential for impact on the climate of the region — Potential for spills of contaminants into water	— Emergency response plans to mitigate impacts — Construction protocol

Source: U.S. Army Corps of Engineers EIS Scope Document.

Table 6.3.   Scope of the EIS: Social Impact

Category	Issues to be Studied	Methodology to be Used
Aesthetic and landscape/ Visual assessment	— Visual impacts to any National Register-eligible site in proximity to any of the final alternatives	
Archaeological	— Any impact on historic districts, buildings, sites or objects, local character and culture, tradition and heritage will be included — Archaeological surveys for final site	— Survey based on previous archaeological and geological investigations — Magnetometer and high-resolution side-scan sonar surveys will be needed to provide electronic data which can be analyzed to assess the potential for any artifacts, such as shipwrecks, followed up by diver reconnaissance where needed — If resources are found which are eligible for listing on the Register of Historic Places, ways to avoid, then minimize, impacts to cultural resources will be considered and discussed. If avoidance is not an option, a Memorandum of Agreement may be required to mitigate potential impacts

(Continued)

Table 6.3. (*Continued*)

Category	Issues to be Studied	Methodology to be Used
Safety issues	— Safety considerations will include public and employee safety through construction, operation and decommissioning	— Design standards for the structures will be explained. List of preparers will include the names and qualifications of persons who were primarily responsible for preparing the EIS and agency personnel who wrote basic components of the EIS or significant background papers must be identified. The EIS should also list the technical editors who reviewed or edited the statements. Cooperating agencies and their role in the EIS will be listed
Public involvement	— List the dates, locations and nature of all public notices, scoping meetings and hearings. The scoping meeting transcripts and summary of comments report to be provided as an appendix	

Source: U.S. Army Corps of Engineers EIS Scope Document.

Table 6.4.   Alternative Sites for the Cape Wind Project

Alternatives	Locations
Shallow water alternatives	1. Horseshoe Shoal (preferred site) 2. Tuckernuck Shoal (off Nantucket Island)
Onshore alternative	3. Massachusetts Military Reservation on Cape Cod

Combined alternative	4. New Bedford Harbor and
	   Horseshoe Shoal could be combined,
	   each with smaller sites combined
	   to achieve the same purpose

Deep water alternative	5. Area south of Tuckernuck Island

Five screening criteria are used to evaluate those alternatives: availability of renewable energy (i.e. wind power classification); ISO New England grid connection availability (connection point, transmission/distribution lines, efficiency/apacity); available land or water area; engineering constraints (constructability, geotechnical conditions, water depths); and legal/regulatory constraints (i.e. endangered species, shipping channels, etc.).

6.5   Public Reaction to Cape Wind

Like any other large-scale engineering system, the Cape Wind project gave rise to different reactions from various groups in the community and in the Commonwealth of Massachusetts. An opinion poll[20] done by the *Cape Cod Times* (openly siding

---

[20] A total of 588 interviews with residents of Barnstable, Nantucket and Dukes County were conducted between February 12–22, 2004, with a margin of error of +/– 4%. The survey was conducted by the Institute for Regional Development at Bridgewater State College and was sponsored by WCAI/WNAN and the *Cape Cod Times*. See http://www.wgbh.org/cainan/article?item_id=1481753 for more information.

with opponents) indicated a nearly even split between opponents and proponents of the project.

Among the opponents, "aesthetics" was the number one reason cited for their opposition. Other reasons included environmental concerns, poor location, wildlife conservation and fishing, as well as objections to private usage of public land for profit. In the following paragraphs, we will be looking at the positions of different groups of stakeholders and decision-makers.

6.5.1 *Project Opponents*

Shortly after the developer filed an application for a data tower in Nantucket Sound, a well-organized and well-financed opposition group called the Alliance to Protect Nantucket Sound formed and voiced its opposition. The Alliance based its opposition on aesthetics as well as concerns about fisheries, tourism, migrating birds, marine mammals and the lack of need for the Cape Wind proposal. Supported by conservation groups such as the Humane Society and the Barnstable Land Trust, the alliance filed a lawsuit against the jurisdiction of the U.S. Army Corps of Engineers to issue a permit for the data tower (Watson and Courtney, 2004). The lawsuit and its appeal were dismissed, with the U.S. Court of Appeals ruling in February 2005 that the Corps had indeed jurisdiction over the project.[21] Still, it is expected that the Alliance will indeed file a lawsuit on the basis of an inadequate environmental impact statement, should the Corps approve Cape Wind's application.

The Alliance has marshaled many other local and state organizations in opposition to the Cape Wind project. Table 6.5 shows groups that have expressed "concerns" about

---

[21] *Electric Utility Week*, "U.S. court affirms Army Corps' jurisdiction over Cape Wind, a big win for developers", February 21, 2005, p. 9.

Table 6.5.   Groups with "Concerns" about the Cape Wind Organization

Actor Groups	Organizations/Entities/Individuals
Political figures	Governor Mitt Romney, MA Attorney General Thomas Reilly, Senator Edward M. Kennedy, Senator Robert O'Leary, Congressman William Delahunt, State Representative Demetrius, Atsalis State Representative Eric Turkington
Towns & local business organizations	Town of Barnstable, Town of Chilmark, Town of Edgartown, Town of Mashpee, Town of Nantucket, Town of Yarmouth, Barnstable County Assembly of Delegates, Cape Cod Chamber of Commerce, Falmouth Chamber of Commerce, Hyannis Area Chamber of Commerce, Martha's Vineyard Chamber of Commerce, Nantucket Chamber of Commerce, Chatham Chamber of Commerce, Harwich Chamber of Commerce, Nantucket Online, Yarmouth Area Chamber of Commerce
Preservation groups	Barnstable Land Trust, Humane Society of the United States, International Wildlife Coalition, Massachusetts Society for the Prevention of Cruelty to Animals, Ocean Conservancy, Pegasus Foundation, Three Bays Preservation, Wampanoag Tribal Council, Save Popponesset Bay
Fishermen's associations	Massachusetts Fishermen's Partnership, Massachusetts Commercial Fishermen's Association, Massachusetts Marine Trades Association, Cape Cod Marine Trades Association, Edgartown Charter Fishing Association, Edgartown Shellfish Organization

(*Continued*)

Table 6.5.   (*Continued*)

Actor Groups	Organizations/Entities/Individuals
Other groups (navigation, aviation, boating, real estate, etc.)	Cape & Islands Harbormasters Association, Hy-Line Cruises, Steamship Authority, Barnstable Municipal Airport Commission, Island Airlines, Nantucket Airport Commission, Martha's Vineyard Airport, Marstons Mills Airport, National Air Traffic Controllers Union, Cape TRACON, Cape Cod & Islands Association of Realtors

Source: Alliance to protect Nantucket sound.

the Cape Wind project, as reflected on the Alliance to Protect Nantucket Sound website.

6.5.2   *Project Proponents*

The project also has its share of proponents, both in Cape Cod as well as in the nest of Massachusetts and the U.S. Table 6.6 shows the supporters of the project as reflected in the Cape Wind (developer) website. Table 6.7 shows groups that see offshore wind power as beneficial, but support a NEPA process to identify environmental and social impact.

6.6   Stakeholder Involvement in the Cape Wind Project

6.6.1   *Scoping Hearings*

In March 2002 the Army Corps of Engineers held two EIS scoping meetings, in Boston and in Yarmouth, Massachusetts. The Corps invited federal, state and local agencies, affected Indian tribes, interested private and public organizations, and individuals to attend the scoping meetings. Stakeholders were also allowed to submit written comments by mail or email to

Table 6.6. Supporters of the Project

Actor Groups	Organizations/Entities/Individuals
Local citizens groups	Clean Power Now, Vineyarders for Clean Power, Islanders for Wind Power (Nantucket)
Environmental organizations	Greenpeace USA, The Coalition for Buzzards Bay, Green Decade Coalition, Buzzards Bay Action Committee, Northeast Sustainable Energy Association, Clean Water Action, Clean Air-Cool Planet, Massachusetts Climate Action Network, Boston Climate Action Network, Religious Witness for the Earth, African American Environmentalist Association, Thompson Island Outward Bound Education Center, Toxics Action Center, Sustainable South Shore, Envirocitizen, Cape & Islands Self-Reliance
Health organizations	American Lung Association — Massachusetts & Maine Chapters, Cape Clean Air, Healthlink, Pilgrim Watch, Citizens Action Network
Business and labor organizations	Cape Cod Area League of Women Voters, American Wind Energy Association, American Council on Renewable Energy, Seafarers International Union, Maritime Trades Council, New England Regional Council of Carpenters, International Brotherhood of Electrical Workers Local 103, International Association of Bridge, Structural, and Ornamental Iron Workers, Industrial Division of the Communications Workers of America Local 201, Cape Cod Internet Council, Coalition for Environmentally Responsible Economies, Mass Energy Consumers Alliance, Competitive Power Coalition of New England

(Continued)

Table 6.6. (*Continued*)

Actor Groups	Organizations/Entities/Individuals
Towns	Town of Truro, Town of Lenox, Town of Westport
Political figures	George D. Bryant (Provincetown Representative), Barnstable County Assembly of Delegates, Donald L. Carcieri (Rhode Island Governor), Daniel E. Bosley (Chair, MA House Committee on Government Regulations), Michael W. Morrissey (Chair, MA Senate Committee on Government Regulations), John J. Binienda (Chair, MA House Committee on Energy), Susan C. Fargo (Chair, MA Senate Committee on Energy), Massachusetts State Representatives (Paul Demakis, Matthew Patrick, Robert Koczera, Frank Smizik, Doug Peterson, James Eldridge, Paul Donato, Patricia Jehlen)

Source: www.capewind.org.

Table 6.7. Environmental Groups that Support the Ongoing Environmental Impact Review Process through the National Environmental Policy Act (NEPA) and the Massachusetts Environmental Policy Act (MEPA)

Organizations Supporting Permitting Process

- Conservation Law Foundation
- MASSPIRG
- Union of Concerned Scientists
- American Rivers
- Friends of the Earth
- Cape & Islands Renewable Energy Collaborative
- Natural Resources Defense Council
- World Wildlife Fund
- People's Power and Light

Source: capewind.org.

the New England District of the U.S. Army Corps of Engineers. Seventy-three stakeholders submitted oral testimonies at these two hearings. Another 120 written testimonials were submitted by regular mail and email.

In addition to the formal hearings, there were informal gatherings convened by the Massachusetts Environmental Policy Act Office, the Cape Cod Commission, the Martha's Vineyard Commission, and the Nantucket Planning and Economic Development Commission (Watson and Courtney, 2004).

6.6.2 *Massachusetts Technology Collaborative Stakeholder Process*

The Massachusetts Technology Collaborative (MTC) is a state agency that administers the Commonwealth's Renewable Energy Trust Fund. The MTC initiated a stakeholder process from October 2002 to March 2003. The goal of the MTC process was

(1) to achieve a better shared understanding of the Cape Wind project's potential benefits and environmental impacts; (2) to shed light on the regulatory process and policy drivers behind the project; (3) to develop a mutual understanding among the conflicting positions of project proponents and opponents; (4) to provide data and information to reveal any areas of factual or philosophical agreement among the stakeholders; and (5) to help prepare all parties to review the material to be presented in the EIS and participate effectively in the regulatory process (Watson and Courtney, 2004).

The existence of the MTC stakeholder process made Cape Wind attractive as an actual case study for SAM-PD. The

MTC stakeholder process was not part of the formal permitting process, but allowed stakeholders and decision-makers to improve their understanding of the contentious issues surrounding the Cape Wind project. For this case study, the MTC stakeholder process served as part of the stakeholder conflict assessment and joint fact-finding steps that form the basis of the SAM-PD process.

Greg Watson, a facilitator with the MTC, has looked at the lessons learned from the MTC stakeholder process. He notes:

> While technical concerns and potential impacts of a single wind farm proposal can be analyzed thoroughly through the NEPA process, even some supporters of the Cape Wind project are troubled by the implications of moving forward absent the kind of publicly vetted structure and compensation environmental advocates have always rightfully demanded for other kinds of energy development projects on public lands. The situation will become more complex as the review begins for other pending offshore projects, including some that are speculative in nature and raise additional concerns. Until a system is established, offshore wind farm developers face tremendous procedural and economic uncertainty (Watson and Courtney, 2004).

6.6.3 *Draft Environmental Impact Statement (DEIS) Hearings*

Four public hearings were held in December 2004 in Oak Bluffs, West Yarmouth, Nantucket and Cambridge, MA. A short narrative on the Cambridge meeting was presented at the beginning of this book. In the meetings in West Yarmouth and Nantucket, most of the comments were against the Cape Wind project, while in the Cambridge meeting supporters outweighed opponents by three to one. For the case study, transcripts of the meetings were used to explore stakeholder inputs for the SAM-PD process.

6.7 Major Sources of Dispute in the DEIS

6.7.1 *Usefulness: Demand for Electricity*

According to the opponents, ISO NE planning documents show that in 2006 the Southeastern Massachusetts region (including Cape Cod and known as "SEMA") will have 3,350 MW of supply and a projected peak demand of just 2,180 MW (APNS, 2003).

6.7.2 *Birds Colliding with Towers*

The Massachusetts Audubon Society is calling for a three-year study to explore seasonal variations in bird migration and area usage. The U.S. Fish & Wildlife service calls the developers' proposal "not sufficient", and recommends at least a three-year study for birds alone. The MA Dept. of Marine Fisheries has expressed "serious concerns centering on the potential risks to migratory birds" (APNS, 2003).

Proponents on the other hand point to the many cases in Europe where bird fatalities are minimal. In the case of inland wind turbines on Mount Waldo, Vermont, a small-scale turbine farm, no bird had been reported killed within a year of its operation (Grady, 2003).

6.7.3 *Impact on Fisheries*

The Mass. Division of Marine Fisheries anticipates "direct negative impacts to fisheries resources and habitat..." (APNS, 2003). Many fishermen's associations have expressed concern over the potential impact on their revenues and commercial fishing populations and believe the DEIS was not sufficiently rigorous in addressing the issue.

6.7.4   *Impact on Marine Mammals*

Opponents such as the Humane Society indicate that the EIS has neglected the impact of construction noise and operating vibrations on marine mammals, particularly the North Atlantic right whale and other whale species.

6.7.5   *Tourism*

Disagreements exist on potential impacts on tourism in the region. The impact of tourism is very important to the region, since more than 21% of the jobs in Cape Cod were in tourism-related industries.

6.7.6   *Lease*

Many residents are concerned with the private use of public land (or waters, in the case of the Cape Wind project), and would like the company to pay royalties as part of a lease. While Cape Wind has essentially agreed to a lease agreement if required by Congress, it contends that the feasibility of off-shore wind energy facilities would be decreased by such lease payments.

6.7.7   *Local and Regional Economic Impact*

In April 2003, Cape Wind released an economic impact report that said the project would create 600 to 1,000 jobs in Massachusetts. During the construction phase, Cape Wind estimated an economic contribution to Massachusetts of $85 million to $137 million annually. They also estimated between $6 and $10 million in increased personal and business tax increases for the state budget. On the other hand, the Alliance published a report in 2003, which stated that without tax credits, the cost of wind energy would be $85/MWh, or twice as high as the $42/MWh for gas-fired plants.[22]

6.7.8 *Aesthetics*

Of course, a major disagreement also exists on the visual impact of the wind farm. While both proponents and opponents agree that the wind farm would be visible from the shore at certain times, they do not agree on the degree of the impact and its characterization. Cape Wind will be painting the turbines such that they blend in with the color of the horizon, but there will still be impact during the night and even during the day. While proponents consider the sight to be beautiful, opponents believe it will destroy the character of Nantucket Sound. There is a consensus among proponents that all the other objections by the opponents stem from a basic aesthetic consideration rather than environmental concerns.

6.8 Chapter Summary

In this chapter we introduced the Cape Wind project and its technical, environmental and social aspects. We looked at the legal context of the case, and the permitting process taking place under the jurisdiction of the U.S. Army Corps of Engineers. Furthermore, we analyzed public positions regarding the Cape Wind project and identified major sources of conflict among stakeholders. In the next chapter,

---

[22] *Electric Utility Week,* "Cape Wind offshore project gets boost from report claiming benefits to economy", April 28, 2003.

we will present the application of the SAM-PD process as an alternative (yet complementary) approach to making decisions for the Cape Wind project, and explore its merits and drawbacks.

Bibliography

APNS (2003) Technical concerns. Alliance to Protect Nantucket Sound, http://www.saveoursound.org/workings/technical.shtml (last accessed March 22, 2005).

Grady, M. (2003) Reaping the wind in a brand new age. *Journal of the Conservation Law Foundation*, 9(2).

Watson, G., Courtney, F. (2004) Nantucket Sound offshore wind stakeholder process. *Boston College Environmental Law Review*, 31: 263–284.

Chapter 7

Stakeholder-Assisted Modeling
of Cape Wind

In this chapter we will explore the stakeholder-assisted modeling and policy design process, as applied to the actual case study of the Cape Wind Offshore Wind Energy Project. The case study is an effort at validating the research hypothesis of this book (revisited below), and will therefore focus heavily on the systems representation aspect of the process, rather than negotiation and implementation issues which are very much ongoing.

7.1 Applying the SAM-PD Process to the Cape Wind Project

As introduced in the first chapter, the argument of this book is that stakeholder-assisted systems representation for an engineering system can produce a *superior* representation compared to traditional, expert-based representations. We defined the criteria for a *superior* representation as follows:

- Inclusion of a plurality of views;
- Usefulness of representation as a thought expander for stakeholders;

- Usefulness of representation for suggesting strategic alternatives for improved long-term management of the system;
- Capturing effects that expert-only representation couldn't capture;
- Completeness of representation (taking into consideration the technical, social, political and economic aspects).

Here, "traditional, expert-based" refers to a system representation that is mainly created by experts at the request of a decision-making entity (e.g. permitting agency) as part of the technical analysis of an engineering system. As with the case of the Cape Wind project, even traditional processes make use of stakeholder inputs in a limited fashion and are therefore not entirely limited to expert analysis. Thus, the main difference between a stakeholder-assisted and an expert-based system representation is the centrality of stakeholder involvement, the degree of involvement and the timing of the involvement.

The Cape Wind project serves as a useful case study to explore the validity of this hypothesis, partly because an expert-based permitting process has been ongoing for more than three years. As described in the previous chapter, the characteristics of the case make it a good example of an engineering system with uncertain social and environmental impact and prone to extensive controversy and stakeholder conflict. This makes the case attractive as an SAM-PD application.

Furthermore, and importantly for this research, the scope of the environmental impact assessment requested by the U.S. Army Corps of Engineers can serve as the expert-based representation for comparison with a stakeholder-assisted representation based on the criteria and measures described in Chapter 1.

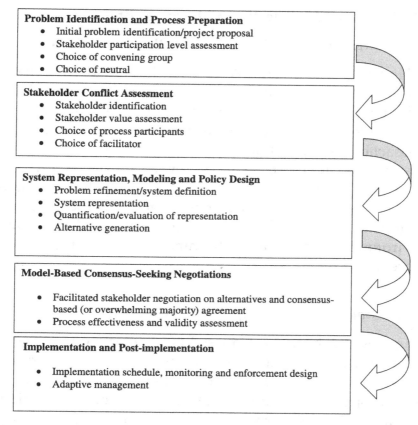

Fig. 7.1. Stakeholder-assisted modeling and policy design (SAM-PD) process diagram.

In applying the SAM-PD process (see Fig. 7.1) to the Cape Wind case study, we emphasize that the actual decision-making process is a traditional permitting process. Still, at this stage of the Cape Wind process, the usefulness of a new collaborative process cannot be ruled out. The Commonwealth of Massachusetts is still fighting to take charge of the permitting process, and if they are successful, the process may need to start from the beginning.

7.2   Problem Identification and Process Preparation

7.2.1   *Initial Problem Identification/Project Proposal*

In November 2001, Cape Wind Associates filed an application for an offshore wind facility in Nantucket Sound. As indicated by the application, the aim of the project was to provide 420 MW of renewable energy to Cape Cod, Massachusetts, as part of the state plan to produce 5% of its energy from renewable sources by 2010. The U.S. Army Corps of Engineers, as the permitting agency, has the mandate to approve the project if it judges that the potential benefits of the project will outweigh its potential social and environmental cost, as determined through the NEPA process.

7.2.2   *Stakeholder Participation Assessment*

Using the PLP heuristic proposed in Chapter 5, one can get a sense of the level of stakeholder participation necessary for the project (Table 7.1). Given its characteristics, the Cape Wind project would require the highest level of participation, which is stakeholder participation in decision-making. While this may not be practical, it is imperative that stakeholders are involved as much as possible in system representation, assessing risks and making recommendations. Therefore, it makes sense to use the SAM-PD process to structure the NEPA analysis.

7.2.3   *Choice of Convening Group*

The convening group has to include organizations that have the mandates, authority and resources to ensure the success and effectiveness of the collaborative process and the power to potentially enforce its agreements. Therefore, the main

Table 7.1.   Participation Level Point (PLP) Heuristic for Cape Wind

The Cape Wind project:

- Has contentious jurisdictional issues
- Affects a wide range of stakeholders
- Has stirred considerable controversy
- Has potentially controversial cost distribution impacts on stakeholders
- Depends on federal tax credits and carbon trading for part of funding
- Involves significant scientific uncertainties in environmental impact
- Has a strong environmental justice component
- Is under the permit jurisdiction of the U.S. Army Corps of Engineers, which has been criticized for its way of handling the permit process
- Has vocal opponents who have threatened the developer with lawsuits
- Involves a variety of scientific and technical agencies (governmental, private sector, academic) who have significant expertise to offer if engaged
- Involves uncertain long-term effects that could be best mitigated through adaptive management
- Has strong obstructionist potentials due to polarized environment

According to the heuristic, these characteristics put the Cape Wind project within Arnstein's "Public Participation in Final Decision" category, which is the highest level of participation.

permitting agency has to be part of the convening group. Other organizations that may be helpful as conveners can be determined by assessing stakeholder bibliography. Suitable candidate organizations for co-convening the process include the Massachusetts Technology Collaborative and the Massachusetts Environmental Protection Act Office. Several stakeholder groups have expressed a desire for the State House and Senate to be involved in managing the process. Table 7.2 shows an ideal convening group, with sufficient authority for the collaborative process to succeed. Such a convening group would reduce the incentive of the Commonwealth of Massachusetts to litigate against the jurisdiction

Table 7.2. **Proposed Convening Group for Cape Wind (Based on Stakeholder Inputs)**

- Federal
  - U.S. Army Corps of Engineers
  - U.S. Environmental Protection Agency
  - U.S. Department of Interior

- Commonwealth of Massachusetts
  - Massachusetts Environmental Protection Act Office
  - Joint Massachusetts Senate-House Appointed Representative Committee
  - Massachusetts Technology Collaborative
  - Energy Facilities Siting Board

of the U.S. Army Corps, and would ensure that state concerns are directly taken into consideration. It would also provide the U.S. Environmental Protection Agency (EPA) and the U.S. Department of the Interior (DOI) with an incentive to invest in generating science through their experts. A U.S. DOI involvement would also allow for lease issues to be integrated into the permitting process.

7.2.4 *Choice of Neutral*

Selecting the neutral is the task of the convening group in its initial sessions. The neutral and his group can be professional mediators, community elders (no obvious choice for the Cape Wind project), or national figures trusted by all sides. Given the controversy of the case, an outside mediator with a track record of neutrality in high-profile cases may be a good choice. While not necessarily a choice for this particular case, former President Jimmy Carter has played this role for other cases of conflict.

7.3   Stakeholder Conflict Assessment

7.3.1   *Stakeholder Identification*

As we discussed in Chapter 6, the MTC stakeholder process for the Cape Wind project identified many of the key stakeholder organizations and provided an opportunity for stakeholders to express their interests and concerns. We used the process transcripts as a starting point for our stakeholder conflict assessment. In addition to the stakeholders identified in the MTC stakeholder process, we also used the U.S. Army Corps of Engineers hearing transcripts to identify further stakeholders and their positions. Tables 7.3–7.5 identify key decision-making stakeholder entities in the Cape Wind project with their respective SPK (stake, power and knowledge) characteristics. Figure 7.2 divides the key stakeholders of the Cape Wind project that we identified into four major categories.

While we did not have access to all of them, we contacted 44 organizations from a total of 62 identified organizations during the stakeholder value assessment stage.

7.3.2   *Stakeholder Value Assessment (Direct)*

From April to November 2004, more than 44 key stakeholder organizations in the Cape Wind controversy were asked to fill out a survey on their perspectives on offshore wind siting criteria and other important considerations. From this sample, 27 organizations (61%) filled out the survey. Five organizations responded that they would not be comfortable providing any feedback, because of their involvement in the permitting process or lack of substantial insight into the project and, despite multiple attempts to contact them, 13 organizations did not respond at all. The purpose of the survey was to elicit the issues, linkages and values stakeholders considered important

Table 7.3.   **Federal Stakeholders in the Cape Wind Project**

Stakeholder organization/Entity	Stake/Power/Knowledge
U.S. Army Corps of Engineers	Current lead permitting agency with federal jurisdiction over Cape Wind. Expertise in navigational safety
U.S. Environmental Protection Agency	Mandated with the Clean Water Act and the Clean Air Act. Expertise in air and water pollution
U.S. Fish and Wildlife Service (DOI)	Mandated with the Endangered Species Act and the Migratory Birds Treaty Act. Expertise on migratory birds, marine ecosystem
National Marine Fisheries Service (DOI)	Mandated with the Marine Mammals Protection Act. Expertise in fisheries, marine ecosystem and marine mammals
National Park Service (DOI)	Mandated with impact on historic places and national parks
U.S. Coast Guard	Mandated with navigational safety, marine search and rescue operations and national security
Federal Aviation Administration	Mandated with air and navigation safety
U.S. Department of Energy	Mandated with the National Energy Act. Expertise in renewable energy and energy economics
Federal Energy Regulatory Commission	Regulates and oversees energy industries in the economic and environmental interest of the public
U.S. Geological Survey (DOI)	Expertise in marine geology
Minerals Management Service (DOI)	Mandated with lease management for public resources. Expertise in offshore leases for oil and gas

(*Continued*)

Table 7.3. (*Continued*)

Stakeholder organization/Entity	Stake/Power/Knowledge
U.S. Department of Commerce	Expertise on local, regional and national economic impact of project
U.S. Legislative Branch (Congress and Senate)	Jurisdiction over new legislation with regards to national energy policy or ocean usage under federal jurisdiction
U.S. Federal Courts and Courts of Appeal	Jurisdiction over lawsuits filed at the federal level with regards to the Cape Wind project

Table 7.4. Commonwealth of Massachusetts Stakeholders in the Cape Wind Project

Stakeholder organization	Stake/Power/Knowledge
Massachusetts Environmental Policy Act Office	State permitting agency supervising the State Environmental Impact Report, with expertise on environmental impact analysis of new project
MA Department of Environmental Protection	402 Water Certification, waterway licensing, air pollution impact
MA Energy Facilities Siting Board	Approval of new power generation processes
MA Division of Marine Fisheries	Fisheries management for areas under state jurisdiction, expertise in marine mammals and organisms
MA Coastal Zone Management Office	Mandated with coastal zone management for the Commonwealth of Massachusetts
MA Division of Fisheries and Wildlife	Equivalent to the USFWS. Mandated with state's protected and endangered species acts. Expertise in fisheries, marine ecosystems and migratory birds

(*Continued*)

Table 7.4. (*Continued*)

Stakeholder organization	Stake/Power/Knowledge
MA Board of Underwater Archaeological Resources	Mandated with protection of underwater heritage sites, historical shipwrecks
MA Historical Commission	Concerned with visual impact to historic areas
Massachusetts Technology Collaborative	Charged with the state's renewable energy fund. Interested in promotion of renewable energy in Massachusetts

Table 7.5. **Local and Regional Government Stakeholders in the Cape Wind Project**

Stakeholder organization	Stake/Power/Knowledge
Cape Cod Commission	Regional planning and regulatory agency supervising a regional land use policy plan for all of Cape Cod
Town of Barnstable	Town directly affected by Cape Wind project
Town of Yarmouth	Town directly affected by Cape Wind project

in the decision-making process. The results were used in conjunction with scoping hearing statements, MTC stakeholder process inputs and stakeholder public statements to help produce a better system representation. The invitation letter and the questionnaire are presented in Appendix A. Table 7.6 shows the participating organizations.

In the following paragraphs, we will present some of the results of the stakeholder survey.

7.3.2.1 *General perceptions of offshore wind energy development*

In the first question of the survey, stakeholders were asked to state their general perception of offshore wind energy

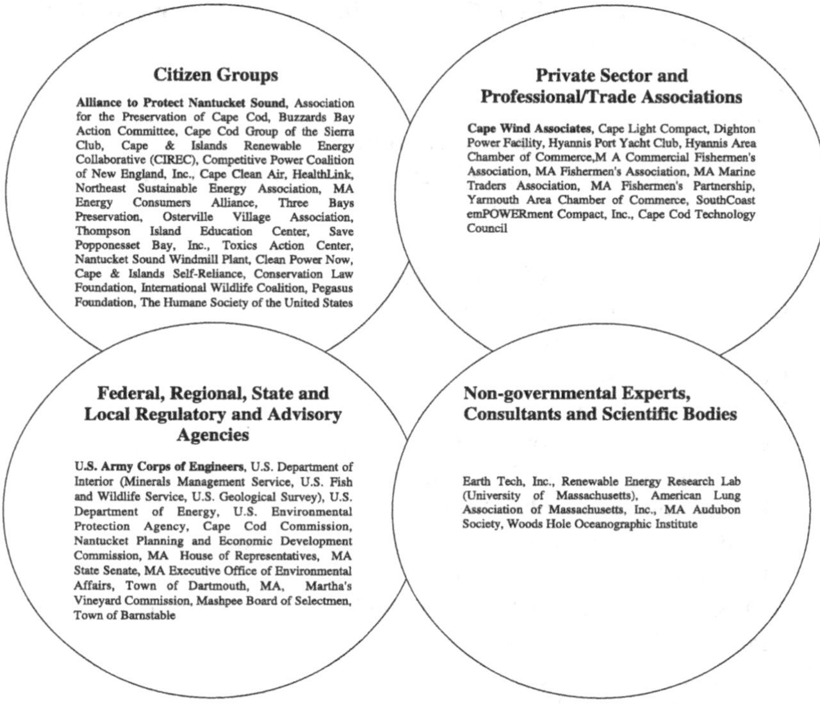

Citizen Groups

Alliance to Protect Nantucket Sound, Association for the Preservation of Cape Cod, Buzzards Bay Action Committee, Cape Cod Group of the Sierra Club, Cape & Islands Renewable Energy Collaborative (CIREC), Competitive Power Coalition of New England, Inc., Cape Clean Air, HealthLink, Northeast Sustainable Energy Association, MA Energy Consumers Alliance, Three Bays Preservation, Osterville Village Association, Thompson Island Education Center, Save Popponesset Bay, Inc., Toxics Action Center, Nantucket Sound Windmill Plant, Clean Power Now, Cape & Islands Self-Reliance, Conservation Law Foundation, International Wildlife Coalition, Pegasus Foundation, The Humane Society of the United States

Private Sector and Professional/Trade Associations

Cape Wind Associates, Cape Light Compact, Dighton Power Facility, Hyannis Port Yacht Club, Hyannis Area Chamber of Commerce,M A Commercial Fishermen's Association, MA Fishermen's Association, MA Marine Traders Association, MA Fishermen's Partnership, Yarmouth Area Chamber of Commerce, SouthCoast emPOWERment Compact, Inc., Cape Cod Technology Council

Federal, Regional, State and Local Regulatory and Advisory Agencies

U.S. Army Corps of Engineers, U.S. Department of Interior (Minerals Management Service, U.S. Fish and Wildlife Service, U.S. Geological Survey), U.S. Department of Energy, U.S. Environmental Protection Agency, Cape Cod Commission, Nantucket Planning and Economic Development Commission, MA House of Representatives, MA State Senate, MA Executive Office of Environmental Affairs, Town of Dartmouth, MA, Martha's Vineyard Commission, Mashpee Board of Selectmen, Town of Barnstable

Non-governmental Experts, Consultants and Scientific Bodies

Earth Tech, Inc., Renewable Energy Research Lab (University of Massachusetts), American Lung Association of Massachusetts, Inc., MA Audubon Society, Woods Hole Oceanographic Institute

Fig. 7.2.   Key stakeholder organizations identified in the stakeholder value assessment stage.

development in the U.S. independent of a specific site. 42% of respondents stated having a very positive view of offshore wind energy development, with an additional 15% expressing a positive view. 19% of respondents expressed a conditionally positive view of offshore wind development, while 11% of respondents expressed skepticism. The remaining respondents (13%) chose not to comment on this question. Furthermore, respondents were given the chance to elaborate on their positions. While not everyone responded, the following comments were made by those who did.

Very positive and positive: Those with "very positive" or "positive" perceptions of offshore wind energy in general

Table 7.6. Stakeholder Organizations Participating in the MIT Stakeholder Survey

Organization
Alliance to Protect Nantucket Sound
Cape & Islands Renewable Energy Collaborative (CIREC)
Cape & Islands Self-Reliance
Cape Clean Air
Cape Cod Chamber of Commerce
Cape Cod Technology Council
Cape Light Compact
Cape Wind Associates
Clean Power Now
Conservation Law Foundation
International Wildlife Coalition
MA Attorney General's Office
MA Board of Underwater Archaeological Resources
MA Division of Marine Fisheries
MA Natural Heritage & Endangered Species Program
Martha's Vineyard Commission
Mashpee Board of Selectmen
Massachusetts Audubon Society
Nantucket Planning & Economic Development Commission
Northeast Sustainable Energy Association
Pegasus Foundation
The Humane Society of the United States
Three Bays Preservation
Town of Yarmouth
U.S. Department of Energy
U.S. Geological Survey
Woods Hole Oceanographic Institute

emphasized the importance of private sector development of renewable energy in realizing the transition from a fossil fuel economy. The current project was identified as a critical element of this transition towards clean and renewable energies (in contrast to other renewable but polluting energy sources). Proponents stated that the many benefits of the project, versus the few drawbacks, made it a compelling choice.

Conditionally positive: Some undecided respondents emphasized the importance of the choice of location and the lack of
proper regulations for offshore wind on the continental shelf as
factors affecting their position. Others stated that the advantages of clean power should be weighed against the drawbacks
of potential environmental impacts, and that they would be
cautious in promoting one or the other position for this reason.

Skeptical: Those "skeptical" of offshore wind energy stated
concerns about the effect of the project on tourism, the environment and aesthetics as reasons for their opposition to the
project. Other concerns included the choice of the site, the
appropriate use of public trust land, cost-benefit analysis,
the lack of appropriate protocols for assessing and mitigating negative impacts, public participation in siting and
mitigation plans in case of problems. Lack of appropriate
federal regulations in dealing with offshore wind energy
development was also stressed. One respondent mentioned
that offshore wind development would not withstand a rigorous cost-benefit analysis. Another view expressed was the
preference for land-based wind farms to offshore wind
farms.

7.3.2.2 *Position on the Cape Wind proposal*

Respondents were asked to state the position of their organization on the Cape Wind proposal. The purpose of this
question was to ensure that the sample was representative of
the different perspectives on the issue, not to represent the
composition of views of all stakeholders in the project. The
choices given were opponent, proponent, undecided, neutral
and no comment. In the respondent sample, 24% identified
themselves as proponents, 20% as opponents, 32% as undecided, 12% as neutral and 8% did not comment on their
position. One respondent who gave a "no comment" answer

Position on Cape Wind

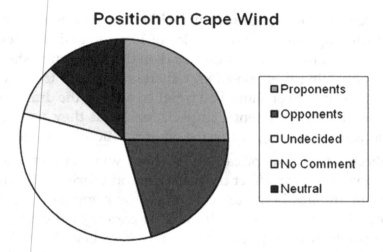

Fig. 7.3. Stakeholder positions on the Cape Wind offshore Wind farm proposal.

objected to the formulation of a position question, indicating that it was improper in the context of this project to summarize positions with a multiple-choice question.

7.3.2.3 *Siting criteria for offshore wind projects*

Stakeholders were asked to consider the process of determining the site for an offshore wind energy project and provide four major criteria they thought were the most important to consider in siting offshore wind energy facilities. Table 7.7 shows the criteria listed by stakeholders based on their stated position on Cape Wind.

7.3.2.4 *Benefits of offshore wind farms*

Stakeholders were asked to identify up to three benefits of offshore wind energy regardless of the site. Table 7.8 shows the benefits identified by stakeholders.

Table 7.7. Stakeholder Siting Criteria Based on Stated Position on Cape Wind

Position	Criteria for siting
Proponents	• Wind resource availability (2) • Commercial feasibility • Community acceptance • Economic externalities • Environmental impacts (2) • Local economic benefits • Nearby shoreline sparsely inhabited • Physical feasibility • Proximity to area of electric demand • Proximity to power transmission grid (4) • Proximity to facilities for construction • Shallow water (less than 50 feet) • Wave heights
Undecided/No Comment/Neutral	• Adequate consideration of alternative sites • Aesthetic considerations (2) • Commercial feasibility • Collaborative process (2) • Local economic benefit (5) • Local economic impacts (4) • Minimal impact on currents/habitats/ecosystems (6) • Mitigation of adverse local impacts • Multiple resource use conflict • Noise impact (construction and operations • Proximity to migratory paths for birds • Proximity to migratory paths for marine life • Proximity to power transmission grid (2) • Ratio of local benefits to impacts • Ratio of power generation to impacts • Shallow water less than 50 feet

(*Continued*)

Table 7.7.   (*Continued*)

Position	Criteria for siting
	• Technical feasibility and adequate wind regime (3)
	• Wave heights
	• Adequate environmental impact information
Opponents	• Aesthetic considerations (3)
	• Alternative sites review (2)
	• Criteria defined by federal, state, local agencies and public
	• Collaborative siting
	• Cost-benefit analysis (4)
	• Decommissioning bonds of outmoded facilities
	• Environmental impact (5)
	• Federal and/or state offshore development regulatory program in place (3)
	• Impact on navigation, aviation (4)
	• Multiple resource use conflict (3)
	• Noise impact (construct. + operation) (3)
	• Proximity to area of electricity demand
	• Proximity to migratory paths for birds
	• Proximity to migratory paths for marine life
	• Potential cumulative impacts of facility over lifetime
	• Sufficient study times

7.3.2.5   *Stakeholder concerns with offshore wind power*

Stakeholders were also asked to identify up to three concerns they might have with offshore wind energy regardless of the site. Table 7.9 shows the concerns listed by stakeholders.

Table 7.8.   Stakeholder Perceptions of Offshore Wind Power Benefits

Position	Benefits of offshore wind
Proponents	• Reduced reliance on imported fossil fuel energy • Greater diversification of region's energy portfolio • Reduced pollutant emissions • New jobs • New economic sector introduced to region • More stable energy prices • Long-term economic benefit to host community • Health benefits of reduced emissions • Cheaper local electricity bills • Hydrogen generation • Implement renewable energy production on large scale • Reduction of greenhouse gases
Undecided/No Comment	• Clean power • More stable energy prices • Economic development within host communities • Reduction of reliance on non-renewable energy • Reduction of greenhouse gases • Monetary benefits to users • Reduction of reliance on foreign oil • Creation of hi-tech jobs • Potential reduction in habitat degradation caused by fossil fuel-based energy production • Potential reduction in human health risk
Opponents	• Lessening fossil fuel dependence • Decreasing air contaminants • Furthering public awareness of alternate energy sources • Renewable energy • More diverse energy portfolio • Energy source with minimal air pollution • Potential economic benefits (currently overstated) • None

Table 7.9.   **Stakeholder Concerns with Offshore Wind Power**

Position	Concerns with offshore wind
Proponents	<ul><li>Fossil fuel energy facing greater competition*</li><li>Trade-off between a project like this and open vista</li><li>Many concerns based on fear will dissipate after construction</li><li>Visual aesthetics (to some people)</li><li>Potential danger for airplanes, boats if not engineered properly</li><li>Navigation</li><li>Boating</li><li>Potential environmental impacts (ocean ecosystem, mainly)</li><li>Private use of and profit from public lands</li><li>Public rejection of wind farms</li></ul>
Undecided/No Comment	<ul><li>Minimal control for local communities bearing adverse impacts</li><li>Questions about ability of host communities to maximize benefits</li><li>Protection of aesthetics (QoL amenity)</li><li>Protection of competing uses (recreation, fishing, etc.)</li><li>Large commitment of a public resource and closure to other use</li><li>Aesthetics</li><li>Environmental impacts (more needs to be known)</li><li>Choice of location</li><li>Contractor ability to get a performance bond</li><li>Exit plan to remove old tired turbines</li><li>Issues of lease</li><li>Potential siting near major migratory routes, harming wildlife</li><li>Anchoring with rip rap introduces artificial reefs (and consequently non-native fauna) and disrupts coastal processes (creating areas</li></ul>

(Continued)

Table 7.9. (*Continued*)

Position	Concerns with Offshore Wind
	of scour and drift) and thus may adversely impact habitat suitability with repercussions for food webs
	• Project sites often chosen on basis of wind availability rather than prior site screening to determine risk averse sites
	• Inadequate study times for impacts
	• Creation of offshore wind power does not necessarily lead to taking other energy sources offline. Overall consumption reduction is key
	• The before and after: construction disturbance (via noise, currents, removal/altering of habitat) and responsibility for the removal of turbines when out of service
	• Appropriation of public land for corporate use (and no current system to regulate/charge for use)
	• There are no guidelines or regulations in place in the U.S. specifically for wind energy plants
Opponents	• Proper licensing
	• Proper environmental review
	• Federal, state and local review
	• Opens door to other forms of offshore development
	• Can endanger valuable offshore resources
	• Cost-benefit analysis doesn't support it from public perspective
	• Choice of location
	• Environmental and economic impacts
	• Decommissioning of out-of-date facilities

7.3.2.6 *Stakeholder proposals for dealing with aesthetics*

With regard to the impact of wind farms on the aesthetics of the region in question, stakeholders were asked to suggest potential objective ways to allow aesthetics to be taken into consideration in the decision-making process. Table 7.10 shows stakeholder suggestions on aesthetics based on their position on the Cape Wind project.

Table 7.10. Stakeholder Suggestions for Dealing with on Aesthetics Impact

Position	Suggestions for aesthetic considerations
Proponents	• "Should be one of very many factors that public agencies evaluate in determining the public interest". • "Analyze sales prices of homes with a wind-farm view, to see if they appreciated less, equal or more than equivalent homes on the coast without a wind-farm view". • "No objective way to take aesthetics into consideration. You could set a price for aesthetics by auctioning off sites (development vs. preservation". • "All ocean views are beautiful. None more or less than any other. Placement in federal waters three miles or more offshore are in the interest of the public good and far enough away that should not be objectionable to the view of reasonable individuals. Would you pit wealthy homeowners vs. working class homeowners view sheds as more or less deserving? What about environmental justice? Should I be compensated for a power plant a mile away whose 500 foot stack I see from my window?"

(Continued)

Table 7.10.   (*Continued*)

Position	Suggestions for aesthetic considerations
	• "I find that these turbines are beautiful, unlike the aging power infrastructure along side our roads that we are so accustomed to that we no longer see it. So, I guess more education on the tradeoffs is needed". • "There's no easy answer to that".
Undecided/No Comment	• Projects may only be granted approval to proceed w/conditions, with aesthetics as one required element in mitigation and with developers/opponents working together to minimize adverse effects • Visualization conducted by objective third party. Trips to see existing facilities, sponsored by objective third party • Move project out of view of the mainland
Opponents	• "Although expense is an issue, siting should be [done as far] away from shore view as possible". • "Mapping of resource areas to show locations of higher and lower impact, with input from affected communities". • "Measure the real visual impacts, lights at night, construction and repair equipment". • "Arguing about aesthetics is not easy to defend, as it is typically seen as being in the interests of the wealthy waterfront property owners, not the public as a whole. I believe aesthetics should be included, but not as a priority of this issue. Rather, the outright appropriation of public lands for private use, the lack of siting review ... and a paucity of regulations regarding offshore wind development are of primary concern".

7.3.2.7 Need for further information for decision-making on offshore wind

Stakeholders were asked to identify areas where more information would be needed for decision-making on offshore wind energy development. Table 7.11 shows the needs

Table 7.11. Further Information Needed for Decision-making[23]

Position	Additional information needed
Proponents	• Analysis of additional potential of offshore wind in case of generous U.S. government support • Economic benefit analysis of electricity production to help understand true economic benefits, effect technology development and labor • Factual information on the enormous benefits of offshore wind power, health benefits from retiring fossil plants
Undecided/No Comment	• Better processes for managing such projects, learning from more successful projects such as the LIPA[24] project. • Assessing whether contractor is financially capable of building a $700 mil project, and can produce a performance bond • Determining areas where wind is optimum and then superimpose wildlife high-use areas such that wind plants are not being proposed for potentially fragile habitats and/or important wildlife use areas • Need to clarify jurisdictional issues regarding permitting and "ownership"

(Continued)

---

[23] It seems this question was poorly phrased. Many stakeholders found this question vague and were unable to provide answers.

[24] Long Island Power Authority (LIPA) Wind Energy Initiative, http://www.lipower.org/cei/wind.html.

Table 7.11.   (*Continued*)

Position	Additional information needed
Opponents	• A pilot government project with a sunset clause to establish independent guidelines • True long-term costs and real impacts • Longer-term studies of bird movement through and residence in the area, as well as all alternate sites • Better understanding of the effect of wind farm construction on currents/ecosystem • Better understanding of impact of noise on marine mammals

identified by stakeholders based on their position on the Cape Wind project.

7.3.2.8   *Selected stakeholder comments on the Cape Wind permitting process*

Stakeholders were also given the opportunity to comment on different aspects of the offshore energy controversy. Some stakeholders focused their responses on the current permitting process and the MTC stakeholder process. Comments in this section are reproduced regardless of position or identity of stakeholders.

- Incentive programs for decreasing fossil fuel usage by the U.S. including, but not limited to, tax credits for citizens who purchase fuel-efficient cars, lights, solar panels, etc. are needed.
- We need to bring on-line a large amount of renewable power in the very near future, and off-shore is where the wind is.
- There is a need to revamp the federal permitting process specifically for off-shore wind projects so that the public feel that they are invested in the process.

- The controversy over Cape Wind might have been avoided had the developer considered the likely controversies/questions prior to putting forward a formal proposal, and chosen a more risk-averse site. For example, the Long Island Power Authority has largely avoided this problem with their wind plant development because there was a pre-proposal site selection process that was intended to minimize abutter concerns and choose sites that were not important habitats for birds. While there are obviously *some* concerns with any project, the LIPA project avoided controversy largely by having done their homework *prior* to choosing a site for development. As testament to this, no conservation group has objected and there have been no serious concerns raised by the U.S. Fish and Wildlife Service or the NY State wildlife department — whereas Cape Wind has been dogged by questions and concerns raised by both the Commonwealth of Mass Division of Fish and Wildlife and the USFWS; both of which have made research recommendations regarding avian habitat use that Cape Wind has neglected to address. I am concerned that their mishandling of this situation will either impede developments in better areas or will allow other developers to believe that they need not address these important concerns prior to finalizing a site proposal.
- The choice of locations is critical. For example, the U.S. Corps of Eng could split up the Cape Wind project into three parts. Two parts in water in different locations and the third on land on the Mass Military Reservation (MMR). For example, the site Horseshoe Shoals is difficult. The U.S. Army Corps of Eng suggests cutting it up into three pieces, one third of which is on land (MMR). The Fed needs to create proper regulations for sea-based

wind. Which fed dept or state dept should be in charge? Royalties need to be paid to the appropriate state.

- Factors such as a federal system to give away public trust land, cost-benefit analysis, protocols for assessing and mitigating negative impacts, public participation in siting and response plans in the case of problems must be in place before permitting begins.

Comments on the MTC process:

- I believe the selection process of the MTC technical advisory system was geared too much toward "undecided" members. Rather, it should have broken off into separate working groups to address various specific issues such as aesthetics, economics and wildlife/habitat concerns, respectively. These working groups could have included folks who may have already formed opinions about the project, but could have provided some very helpful expertise to the review process.

7.3.3 *Stakeholder Value Assessment (Indirect)*

Newspapers and websites: In many cases it is difficult to reach all the key stakeholders for commenting. But there are some indirect ways of considering the views of stakeholders on the system. In the case of Cape Wind we used more than 60 newspaper articles, 47 press releases, and stakeholder websites for information on their values, positions, interests and concerns. An example of one newspaper article used in value assessment is shown below.

> ... The potential for **bird deaths** promises to become a major question surrounding the wind farm ... in an area heavily traveled by **sea ducks, migratory birds**, and some endangered species ... federal wildlife authorities have complained that two widely disparate numbers suggest

that there is **not enough data** to know **how many birds fly through the area**, let alone how many will be killed by the turbines ... One figure used in a version of the report, compiled by **aerial and boat surveys**, counts 210 individual birds over about 180 hours of observation that fly in the zone the turbine blades will sweep. Another figure, based on two months of radar data, shows that 127,697 birds or bats flew through the same area, some of which may have been counted multiple times ...." There is no data for 10 months of the year with the radar survey, and it's clear the other information is **unreliable**", said Vernon Lang, assistant supervisor of the New England Field Office of the US Fish and Wildlife Service .... He says the lack of fundamental data means "we only get a glimpse of what is happening. We need more information".... [Cape Wind says that] the two different bird counts are not designed to be compared, but rather to help Army Corps officials make an informed decision about bird **behavior, habitat, and potential harm** in the area where the wind farm would be built. He said the radar study was taken during **fall and spring migratory months in 2002**, giving a snapshot of the busiest times of the year for birds in Nantucket Sound.

Supporters of the project say that any bird deaths would be minimal **compared** with the millions of birds that die each year colliding with **skyscrapers and cellphone towers**. Opponents, meanwhile, have said any bird deaths are unacceptable (Beth Daley, "Report on possible risks from wind farm fuels ire, *Boston Globe*, October 17, 2004).

The bolded terms contain stakeholder views on performance metrics for the system, views on data issues and measurement methods, reliability and uncertainty of baseline information, metrics for determining good system performance, and views on acceptable risk.

Table 7.12   **Selected Stakeholder Websites for Cape Wind Project**

Stakeholder group	Role	URL (Web Address)
U.S. Army Corps of Engineers	Decision-makers	http://www.nae.usace.army.mil/projects/ma/ccwf/windfarm.htm
Cape Wind Associates	Developer	http://www.capewind.org/index.htm
Alliance to Protect Nantucket Sound	Primary opponents of Cape Wind	http://www.saveoursound.org/
Clean Power Now	Grassroots citizens' group for the Cape Wind project	http://www.cleanpowernow.org/
Windstop.org	Grassroots citizens' group against the Cape Wind project	http://www.windstop.org

Most organized stakeholder groups in the U.S. and other developed countries also have some of their views presented on their websites (see Table 7.12). These are usually far more comprehensive than those that can be found in newspaper articles, but similar to newspaper articles, websites only represent the voice of those already vocal.

Formal and informal hearing transcripts: In many cases, formal or informal hearings are held at different stages of decision-making. Transcripts of these hearings, when available, can shed further light on stakeholder views on the system. EIS scoping meetings were held in March 2002 and DEIS comment meetings were held in December 2004. We used transcripts from these meetings to further inform the assessment of stakeholder values and concerns. Table 7.13 shows the components and linkages identified by stakeholders as important. The numbers in parentheses show the number of stakeholders identifying each of these issues in their comments.

Table 7.13.   Stakeholder Comments in Public Hearings

Category of comment	Issue/Linkage/Component
Project goal and justification	Power available to NE (5)
	Location (5)
	Scale (5)
	Magnitude (5)
	Energy demand assessment (21)
Analysis of alternatives	Technical feasibility (62)
	Economic feasibility (62)
	Comparison with other sources (62)
	Comparison with on-Shore (62)
	Location (62)
	Size/scale (62)
	Power demand assessment. (62)
	Collaborative assessment (1)
	Clarity of assumptions in analysis (62)
Permitting process	Transparency of process (10)
	Need to refine NEPA process for offshore wind energy (10)
	Inclusion of fishermen in planning (55)
	Inclusion of state agencies in planning (55)
Jurisdiction and authority	Use of federal and state resources for profit (26)
	No existing regulations for permitting wind farms (26)
	Compensation of federal government (2)

(Continued)

Table 7.13. (*Continued*)

Category of comment	Issue/Linkage/Component
Energy source	Renewable energy capacity (52)
	Energy efficiency (52)
	Energy costs (52)
	Impact on Air Pollution (52)
Fuel diversity	Diverse energy portfolio (28)
	Dependence on foreign oil (28)
	National security (8)
	Local independence in energy (10)
Economic analysis	Total cost assessment (6)
	Value of savings (6)
	Market value of wind power (6)
	Cost relative to other sources (6)
Wind technology	Impact of emerging technology on feasibility of alternatives (16)
	Assessment of European experience (24)
	Credibility and reliability as a power source (24)
Electricity rate change	Impact on electricity rates in NE (17)
	Reliability issues for wind and other fuels (17)
	Price volatility of fossil fuels (17)
	Potential supply limitations (17)
Fiscal impacts	Potential sale of power to grid (17)
	Share of renewable energy in Cape (17)
	Fiscal impacts on towns (17)
	Subsidies (17)
Environmental trade-offs	Air pollution benefits (11)
	Avoided impacts (11)
	Assessment of emission reduction (9)
	Amount of fossil fuel saved (4)
	Impact on fuel consumption (6)
	Impact on recreational fish (34)
	Impact on commercial fish (34)

(*Continued*)

Table 7.13.   (Continued)

Category of comment	Issue/Linkage/Component
	Vibrations (34)
	Sound (34)
	Sediment Transport (34)
	Creation of unique habitats under towers (10)
	Avian Collision (27)
	Proximity to migration paths (27)
	Loss of habitat (27)
	Loss of marine feedstock (27)
	Bird nesting on structures (27)
Socioeconomic impact	Regional economic impact (5)
	Negative visual impact (41)
	Value of natural resources in area (5)
	Characterization and referenced judgment of visual impact (78)
	Impact on real estate prices (8)
	Impact on local tax revenue (8)
	Impact on recreation (47)
	Impact on tourism (47)
	Impact on fishing (47)
	Impacts on fishing (55)
	Safety in case of adverse weather (55)
	Identification of existing archaeological sites (3)
	Oversight by Mass Historical Soc (3)
	Educational opportunities (7)
	Positive impact on tourism (7)
	Number of local jobs created (13)
	Duration of jobs (13)
	Impact on long-term sustainability (13)
Health and safety impact	Impact on human health (4)
	Impact on local air quality (4)
	Impact on aviation (9)
	Landfall impacts on residential areas (10)
	Landfall impacts on wetlands (10)

(Contionued)

Table 7.13.   (*Continued*)

Category of comment	Issue/Linkage/Component
Construction and removal operations	Construction protocols
	Waste management during construction (2)
	Mitigation measures during construction (2)
	Installation, maintenance, and removal protocols (5)
	Tower and cable maintenance schedule (12)
	Contingency strategy in case of project failure (21)

Note: Extracted from the Stakeholder Comment Summary Transcript from the New England District through April 24, 2002, as part of Section 10 Permit Application Draft Environmental Impact Statement Cape Wind Associates, LLC. Prepared for: US Army Corps of Engineers File No. 200102913.

7.4   Problem Refinement and System Definition

Based on the information gathered in the stakeholder value assessment stage, the Cape Wind project has the following characteristics:

- *System goal*: Providing renewable wind energy for Cape and Islands through installation of a wind farm offshore.
- *Geographical scale*: The area in question includes Nantucket Sound and adjacent land and sea areas (Cape and Islands). However, the general impacts and consequences of the project extend to the region and the nation.
- *Temporal Scale*: The project was proposed in 2001 and its anticipated lifetime is at least 25 years from the time of installation (currently 2006 or 2007). Project lifetime could be extended if turbines, cables and foundations are replaced and/or well-maintained.

- *Institutional issues*:

  o Large-scale disputes among stakeholders in locality and on a nation-wide basis.

o Likely to serve as precedent of offshore wind development in the U.S.
o Lack of existing regulations specifically for offshore wind energy development and ambiguous jurisdictional issues.

7.5   Initial Stakeholder-Assisted Representation

With the stakeholder inputs in the value assessment stage, we created an initial system representation that would serve as a basis for the stakeholder-assisted representation process.

7.5.1   *Identifying Major Subsystems*

Using the CLIOS process approach, the system representation is developed using a top-down approach, with information being categorized in major subsystems. This can be done using the summary that was prepared in the stakeholder value assessment stage. Subsystems can be developed based on their functions or on their technologies. Based on stakeholder inputs, the following subsystems can be initially determined: energy subsystem, socioeconomic impact subsystem, environmental subsystem, and navigation, aviation and safety subsystem.

7.5.2   *Populating the Physical Subsystems*

Based on the stakeholder identification and consultation step, the physical domain can be populated with their major components. Figures 7.5–7.8 show the populated subsystem diagrams for Cape Wind.

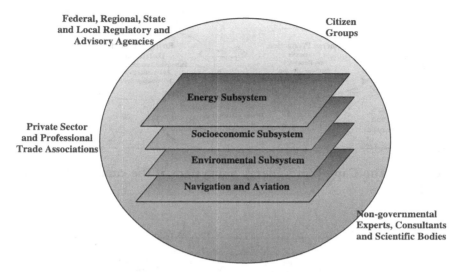

Fig. 7.4. High-level CLIOS system representation of the Cape Wind project. The sphere around the physical systems is the institutional sphere with the major actor groups identified. Here the concept of nested complexity is shown by the complex physical system embedded in institutional complexity.

Energy subsystem: The energy subsystem in the Cape Wind CLIOS consists of four major layers: energy production, energy demand and distribution, energy economics and wind farm characteristics (Fig. 7.5). The dashed arrows show the linkages of the proposed wind farm with the existing layers. Given that the wind farm is currently not in existence, these links do not yet exist.

Socioeconomic Subsystem: The socioeconomic subsystem in the Cape Wind CLIOS consists of four major layers: local and regional economic impact, national impact, social impact and wind farm characteristics (Fig. 7.6).

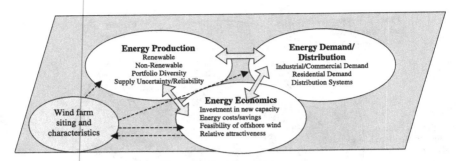

Fig. 7.5. The Cape Wind energy subsystem at the component group level.

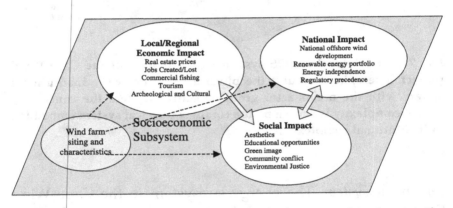

Fig. 7.6. The Cape Wind socioeconomic impact subsystem at the component group level.

Environmental subsystem: The environmental subsystem in the Cape Wind CLIOS consists of four major layers: marine life impacts, avian impacts, air pollution impact and wind farm characteristics (Fig. 7.7).

Navigation, aviation and safety subsystem: The navigation, aviation and safety subsystem in the Cape Wind CLIOS consists of four major layers: navigation impacts, aviation impacts, construction impacts and wind farm characteristics (Fig. 7.8).

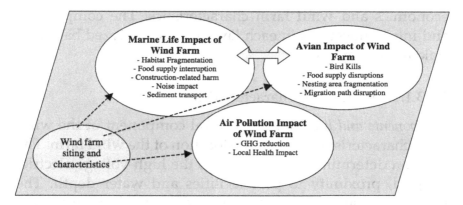

Fig. 7.7. The Cape Wind environmental impact subsystem at the component group level.

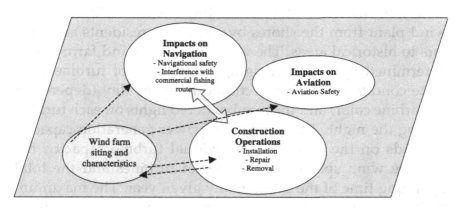

Fig. 7.8. The Cape Wind navigation, aviation and safety subsystem at the component group level.

7.5.3 *Describing Linkages and Components in the Physical System*

7.5.3.1 *Energy subsystem*

The energy subsystem consists of these layers: electricity production, electricity consumption and distribution, energy economics and wind farm characteristics. The components and interconnections in each layer can be analyzed based on stakeholder inputs.

7.5.3.1.1   Wind farm characteristics

Components and linkages: The central component of the wind farm characteristics layer is the location of the wind farm. The location determines the distance to the high-voltage electricity grid, proximity to port facilities and water depth. The technical feasibility of the site depends on these three components. The location also determines the average wind speed that can be expected throughout the year. In addition, the location determines the distance to the nearest residential areas and historical sites, which impacts the visibility of the wind plant from the shores by Cape Cod residents and visitors to historical areas. The visibility of the wind farm is also determined by turbine height, the number of turbines, climate conditions, light reflection during the day (depending on turbine color) and the number of red lights on each turbine during the night. The wind farm power generation capacity depends on the maximum individual turbine capacity, the average wind speed, the number of turbines and the total operating time of the plant in any given year. The maximum turbine generation capacity depends on available wind technology and can change over time as better turbines become available. In summary, we can show the individual components and interconnections for wind farm generation capacity and visibility of wind farm in a causal tree form as illustrated in Figs. 7.9 and 7.10.

Characterizing the components and links: Figure 7.11 shows the completed representation for the wind farm characteristics layer. The location of the wind farm, the number of turbines, their height, maximum generation capacity and color are

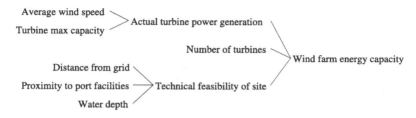

Fig. 7.9.   Components and linkages affecting wind farm energy capacity.

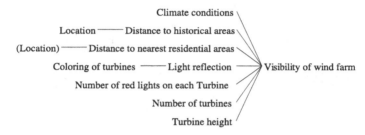

Fig. 7.10.   Components and linkages affecting visibility of wind farm.

open choices. As such they are considered *policy levers*. As discussed in Chapter 2, in a CLIOS representation, we show policy levers with rectangles.

Since location and climate conditions are common drivers between different layers and even different subsystems, they

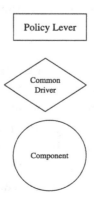

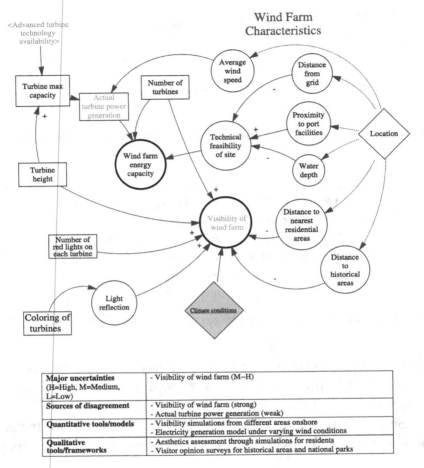

Fig. 7.11. System representation of the wind farm characteristics layer.

are shown as diamonds. Shading indicates that the driver is external to the system, meaning that it is beyond the boundary of the system. The thicker boundaries indicate that a component is considered to be a performance metric. That is, the stakeholders have indicated that the performance of the system depends on how these components fare with regards to the different strategies and alternatives available. In this layer, wind farm energy generation capacity and visibility are the two performance metrics. Components used in the same subsystem but in different layers are shown with "< >" boundaries.

The representation is accompanied by a table that highlights major uncertainties in the layer, sources of disagreement (often on the same uncertainties) and methodologies/tools that can be used to evaluate the layer.

7.5.3.1.2   Energy production

Components and linkages: The electricity production for the Cape and Islands comes from a variety of sources, including existing nuclear energy facilities, gas- and oil fired plants, coal-fired plants and renewable sources, of which the offshore wind energy farm would become one. Oil supply reliability and price fluctuations can impact fossil-based production. The renewable energy production capacity of Cape Cod would depend on sources such as solar energy, onshore wind energy, the proposed wind farm and biomass. The total contribution of the wind farm depends on the total operating hours of the wind farm and average wind speeds. If the wind farm is shut down during the high migration season for some time, the total production capacity can decrease, but this could reduce the environmental impact dramatically. We will look at this link further in the environmental subsystem.

Characterizing components and links: Figure 7.12 shows the completed system representation of the energy production layer. We can determine how to produce the necessary energy when there is a capacity gap (more demand than production). For instance, we can decide whether to invest more in fossil-based energy production or renewable energy. As such, these components are *policy levers*. Here we are concerned with the total amount of renewable energy produced, since the overall goal is for the state to produce as much as 5% of the energy

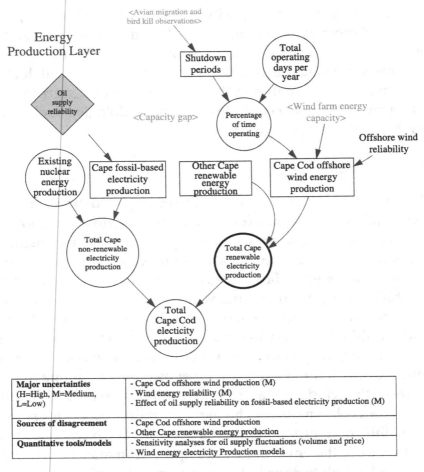

Fig. 7.12.   System representation of the energy production layer.

from renewable sources by 2010 (see Chapter 6). Therefore, total renewable energy production is a performance metric and shown with a thicker border.

7.5.3.1.3   Energy consumption

Components and linkages: An increase in the Cape Cod population and economic growth, along with an increase in per capita consumption in electricity will lead to an increase in residential, business and industrial electricity demand in the Cape and Islands. If the total electricity demand of the region increases beyond the combined renewable and non-renewable electricity production shown in Section 7.5.3.1.2, there will be a capacity gap that has to be overcome either by building additional fossil-based power plants, increasing renewable energy capacity in the region or importing electricity from outside the region from the national grid. Figure 7.13 shows these relationships individually.

Characterizing components and links: Population, economic growth and per capita electricity consumption are external drivers (shown as shaded diamonds). Stakeholders have identified total Cape Cod electricity demand as a performance metric. Figure 7.14 shows the completed system representation for the energy consumption layer.

7.5.3.1.4   Energy economics

Components and linkages: The sum of total annual costs of capital and operation for Cape Wind in addition to the transmission cost make up the basic costs for Cape Wind. However, should we consider additional mitigation and research funds such as environmental research funds for post-construction, federal lease payments, impact mitigation funds and removal and upgrade funds, the cost of electricity produced would be

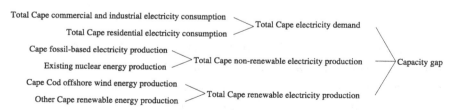

Fig. 7.13.   Factors affecting power generation capacity gap.

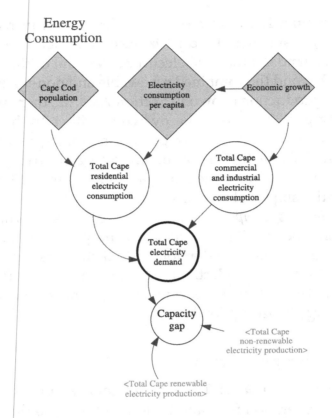

Fig. 7.14. System representation of the energy consumption layer.

higher. How much each of these funds should be, or whether they should exist at all is a matter of negotiation among stakeholders. Figure 7.15 basically shows the impact of different factors on the cost per kWh of electricity.

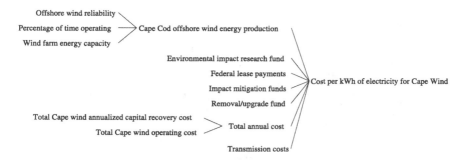

Fig. 7.15. Components making up the kWh electricity cost of Cape Wind.

Characterization of components and links: Figure 7.16 shows the comprehensive system representation for the energy economics layer based on the intermediate step of Fig. 7.15. Here grid transmission costs, oil and natural gas fluctuations and the market value of green credit have been taken as external common drivers (shown as shaded diamonds). Impact mitigation funds, environmental research funds, federal lease payments, removal/upgrade funds and federal tax credits are taken as policy levers (shown as rectangles). Electricity rates for Cape Cod residents and net annual savings for Cape Cod residents are considered performance metrics in this layer.

This ends our representation of the energy subsystem. We then proceed to the environment subsystem of the Cape Wind system.

7.5.3.2 *Environment subsystem*

The environmental impact subsystem includes layers for air pollution impact, marine and avian impacts and endangered species impact.

Energy
Economics

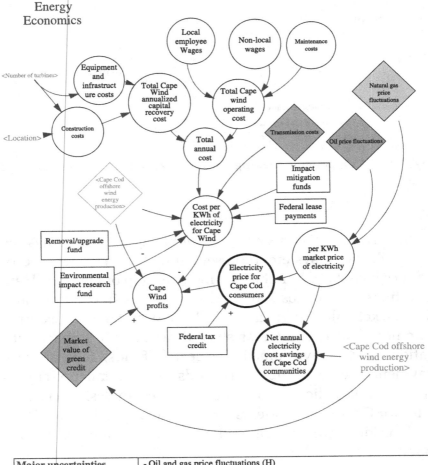

Major uncertainties (H=High, M=Medium, L=Low)	- Oil and gas price fluctuations (H) - Value of green Credit (M)
Sources of disagreement	- Net annual savings for cape cod residents (strong)
Quantitative tools/models	- Energy cost models with cross-fuel-source elasticity

Fig. 7.16.   **System representation of the energy economy layer.**

7.5.3.2.1   Air pollution impact

Components and Linkages: Since the wind farm itself has no air pollution impacts during operation, in this context the air pollution impacts are in terms of avoided emissions. Assuming that the generation capacity of the wind farm could substitute that of coal- or gas-fired power plants, we would have a reduction (compared to the same power generated by fossil-based plants) in greenhouse gases (primarily CO_2), particulate matter (PM), SO_2 and hydrocarbons. The avoided emissions vary based on whether a gas-fired plant or a coal-fired plant is substituted. In addition to general climate change benefits, there are direct health benefits in the form of reduced asthma and other respiratory problems. The reductions can result in lower morbidity (cases of illness) or mortality (deaths). The reductions translate into reduced productivity loss and human life loss, as well as reduced healthcare costs for the region. Figure 7.17 shows the completed system representation of the air pollution impact layer.

7.5.3.2.2 Avian and marine impact

Components and links: The location determines the proximity to avian migration routes and avian seasonal residences. Locations that have a higher number of resident birds in the area have a higher likelihood of bird kills through collision with the wind turbines. In addition, turbine height and the number of lights on the turbines can potentially contribute to bird kills. Shutdown periods during the peak migration seasons can reduce the risk of migratory bird kills significantly, and can serve as a strategy to reduce the environmental impact of the wind farm.

The number of turbines also affects the potential number of bird kills and impact on marine life. When in operation, the vibrations of the turbines may result in substrate movement and fouling, as well as extensive habitat fragmentation. This can in the longer term affect benthos (seabed organisms), commercial fish stocks, marine mammals and prey availability for

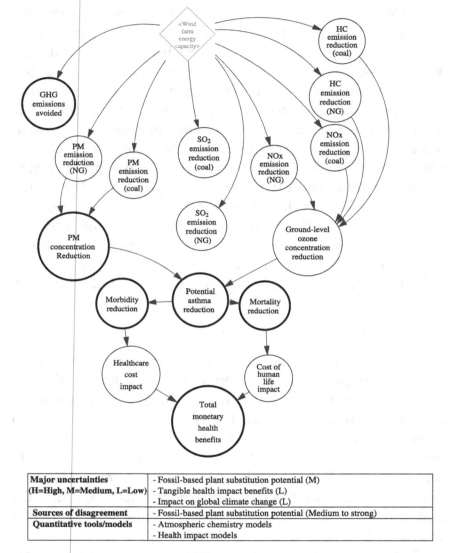

Major uncertainties (H=High, M=Medium, L=Low)	- Fossil-based plant substitution potential (M)
	- Tangible health impact benefits (L)
	- Impact on global climate change (L)
Sources of disagreement	- Fossil-based plant substitution potential (Medium to strong)
Quantitative tools/models	- Atmospheric chemistry models
	- Health impact models

Fig. 7.17.   System representation of the air pollution impact layer.

birds. The potential disruption in the food supply for birds can
result in deaths unrelated to collision with the towers. In addition to the operation of turbines, there is the potential impact
of construction noise on marine mammals. Construction

protocols can address many of these issues by designing ways to reduce noise and habitat destruction during the construction of seabed foundations and installation of the turbines. The existence of an oil tank on the electric transformer platform introduces the risk of water pollution through oil leak. Double-hulled containers can help address this problem.

Component and link characterization: Figure 7.18 shows the system representation of the marine and avian impact layer. Bird

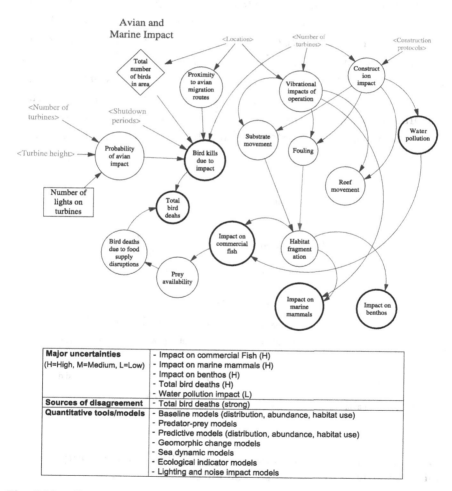

Fig. 7.18.   System representation of the marine and avian impact layer.

kills through collision and total bird deaths due to collision and food disruption, impact on marine mammals, impact on benthos and impact on commercial fish are considered performance metrics in this layer. The number of lights on the turbines is a policy lever, and the number of birds in seasonal residence in the area is an external driver.

7.5.3.2.3 Protected/endangered species

Components and links: Proximity to migration routes affects potential impacts on birds under the Migratory Bird Treaty Act, as well as sea ducks and roseate terns (all protected/endangered species on the state/federal lists). Construction noise can affect marine mammals such as the North Atlantic right whale and other whale species as well as grey seals. Habitat fragmentation can also have an impact on green sea turtles, loggerhead turtles and leatherback turtles. Figure 7.19 shows these relationships.

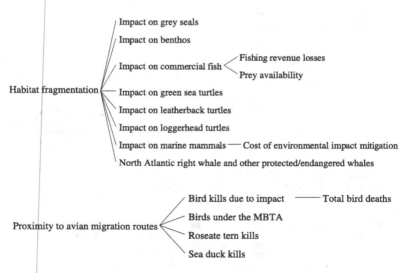

Fig. 7.19. **The impact of habitat fragmentation and proximity to avian migration routes and seasonal residences.**

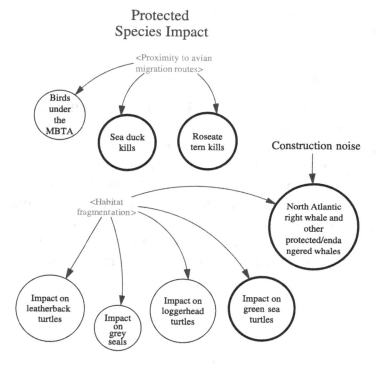

Fig. 7.20.   **System representation of endangered/protected species impact layer.**

Characterization of components and links: Figure 7.20 shows the system representation of the endangered species layer. Impacts on sea ducks, roseate terns, green sea turtles and North Atlantic right whales are particularly

important to stakeholders and can be considered performance metrics.

7.5.3.2.4   Post-construction monitoring

Components and linkages: Stakeholders also emphasized the
importance of post-construction and operation monitoring of
the environmental impacts of the project. These could be
funded by an environmental impact research fund, discussed
in the energy economics layer. The observed impacts could
then result in changes in the way the project is managed.
Changes could include longer shutdown periods, removal of
high-impact turbines, changes in the color and lights or
changes in the number of turbines operating. Figure 7.21

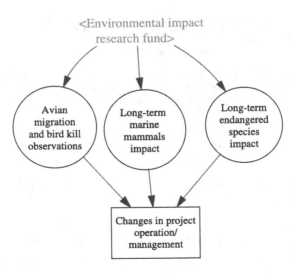

**Fig. 7.21. System representation of post-construction monitoring
layer.**

shows the system representation of the post-construction monitoring. In the next section, we will look at the socioeconomic impact subsystem.

7.5.3.3 *Socioeconomic impact*

7.5.3.3.1 Local and regional economic impact

Components and links: The local economic benefits of the project are summarized in Fig. 7.22. The local costs are summarized in Fig. 7.23. The net local economic benefit is determined by subtracting the local economic costs from the benefits.

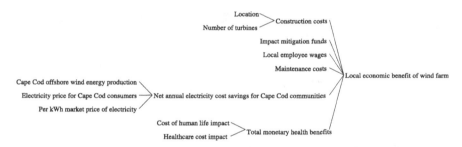

Fig. 7.22.   Local economic benefits of the wind energy project.

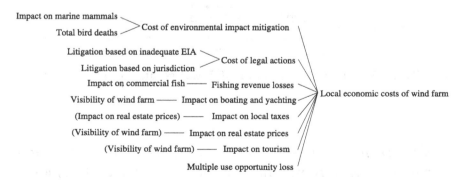

Fig. 7.23.   Local economic costs of the wind energy project.

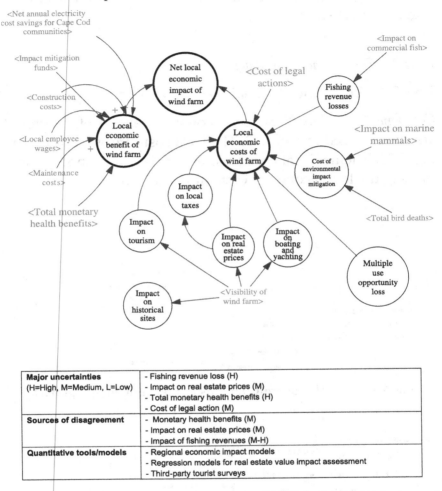

Fig. 7.24. System representation of local/regional economic impact layer.

Characterizing components and links: Figure 7.24 is a comprehensive system representation of the local and regional economic impact of the project.

7.5.3.3.2   Social and cultural impact

Components and links: Stakeholders have expressed concern about the impact of wind farm visibility on the historical character and open vista of Nantucket Sound. On the other hand, some stakeholders see a potential for educational opportunities in renewable energy for schools in the area. Figure 7.25 shows the representation of the social and cultural impacts of the project.

7.5.3.3.3   Stakeholder process

Components and links: Stakeholders commented on the importance of the decision-making process and its impacts on the

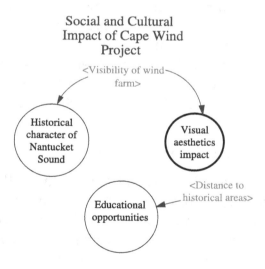

Major uncertainties (H=High, M,=Medium, L=Low)	-Visibility of wind farm (M)
Sources of disagreement	- Impact on historical character of Nantucket Sound (weak) - Impact on aesthetics (strong)
Quantitative tools/models	-Third-party visual simulations and third-party opinion surveys

Fig. 7.25.   System representation of social and cultural impact layer.

validity of the final decision. Stakeholders indicated that they needed to have a more active role in the decision-making. Some also indicated that state agencies should be more actively involved. This would result in a more credible EIS. Stakeholders believed that community acceptance and support was crucial to the success of the project. Involving stakeholders may reduce conflict and the probability of litigation, which may reduce the costs of project implementation delay and potential litigation. Together, these make up the cost of conflict, which tie into the local economic costs of the wind project. Other stakeholders have expressed doubt that stakeholder involvement will actually help in reducing the conflict, as it takes only one unhappy stakeholder to file a lawsuit. The cost of stakeholder involvement and time constraints in decision-making are seen as arguments that weaken the attractiveness of stakeholder involvement. Figure 7.26 shows the system representation of the stakeholder process layer. The major uncertainties seem to lie in assessing the value of the stakeholder process. The qualitative and speculative nature of this layer makes it only useful as a contextual consideration for most stakeholders.

7.5.3.3.4   Long-term impacts of the project

Components and links: Most stakeholders agreed that the Cape Wind project had a precedent-setting nature with extensive consequences that would go beyond the current project. Stakeholders agreed that an approval of the permit would spur new offshore wind energy developments and that its defeat would have serious consequences for future similar developments. An approval would potentially lead to more renewable energy for the Cape and Islands, as well as the United States as a whole. On the one hand, this can have positive long-term impacts on emission reduction, job creation

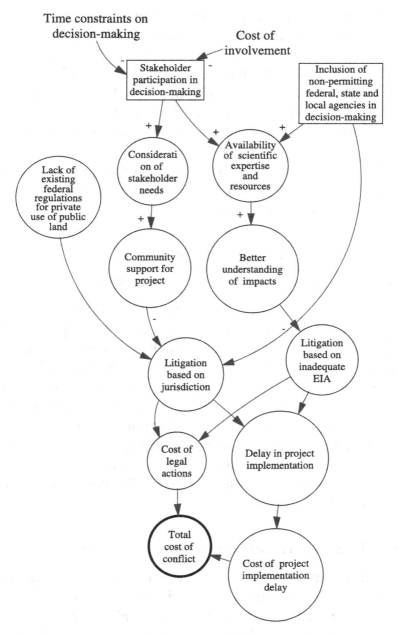

Fig. 7.26.   System representation of the stakeholder process layer.

and health benefits. On the other hand, more wind farms would also mean increased cumulative impact, which may not be linear. For instance, were the entire Eastern seashore covered with wind farms, it would be impossible for birds and marine mammals to find suitable habitats at all. Again this layer is rather speculative and can be used only as a context. Figure 7.27 shows the long-term impacts of the project beyond the current process.

This completes the representation of the socioeconomic subsystem. In the next section we will look at the navigation, aviation and safety subsystem.

7.5.3.4   *Navigation, aviation and safety*

7.5.3.4.1   Navigation and aviation and safety

Components and linkages: The location of the project determines the proximity to navigation routes, impacting the probability of vessel collisions with the wind turbines in adverse weather conditions. Additional factors include the number of turbines, their structural safety and construction protocols for cable restraint design as well as turbine maintenance. Factors affecting the risk of airplane collision with the wind turbines are the number of turbines, the number of lights on turbines, turbine height, proximity to aviation routes and climate conditions. Figure 7.28 shows the navigation and aviation layer.

7.5.3.4.2   Safety, construction and maintenance

Components and linkages: Turbine repair operations depend on the funds allocated to repair and maintenance, as well as the maintenance schedule. The potential of upgrading the turbines with more advanced technologies depends on the availability of funds for upgrading and replacement of

Long-Term Impacts
of Permit

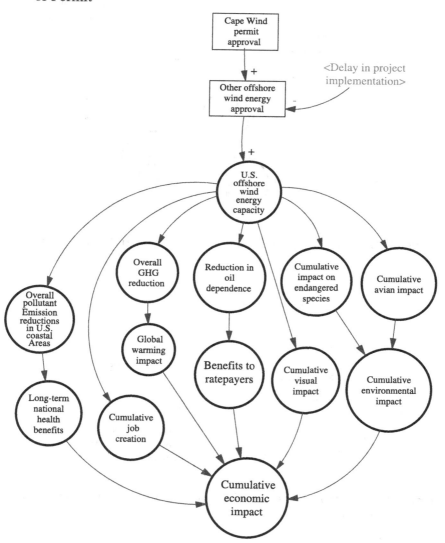

Fig. 7.27. System representation of long-term impacts of project.

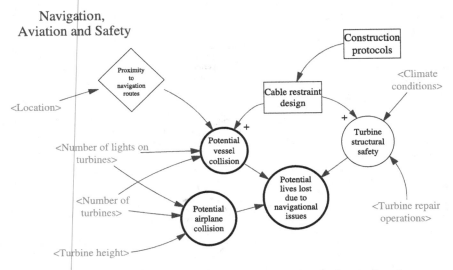

Fig. 7.28.   System representation of navigation and aviation layer.

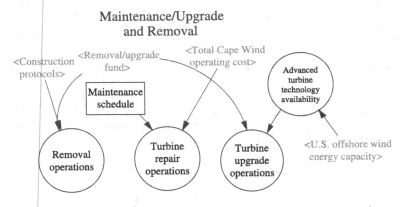

Fig. 7.29.   System representation of the safety, construction and maintenance layer.

turbines. The development of advanced wind technology, on the other hand, depends on the market for wind energy in the U.S., which in turn is highly dependent on the permit of the project, as discussed in the long-term impact layer. A better turbine technology means a higher power generation capacity at a lower environmental impact. Also important for stakeholders is the availability of removal funds to ensure the turbines are adequately removed at the end of the lifetime of the wind farm.

With the last subsystem represented, the initial system representation for the Cape Wind system is complete. In the next section, we will look at the stakeholder refinement and validation of the system representation.

7.6 Stakeholder-Refined System Representation

7.6.1 *Workshop Invitation and Process Preparation*

Once an initial system representation based on stakeholder inputs was developed, the representation was put on the project website.[25] This was done both in the form of a PowerPoint presentation, as well as the actual Vensim model file (with the option to download the software for free). Additionally, the survey results were provided in report format to stakeholders. Figures 7.30 and 7.31 show snapshots of the website.

Stakeholders were then invited to attend a collaborative systems representation workshop at MIT. Efforts for finding a neutral venue on the Cape itself had failed in the weeks prior to the workshop. Most organizations considered the Cape Wind project as a "political hot potato", and did not wish to be co-conveners of the workshop.

---

[25] A complete website for the Cape Wind case study was created to facilitate communication with stakeholders. In addition to the online survey, many stakeholders have provided feedback through the website. The entire process was made transparent for those stakeholders not participating directly in the process. The website can be accessed at http://web.mit.edu/amostash/www/Capewind.

Massachusetts Institute of Technology

Stakeholder-Assisted Modeling and Policy Design Process (SAM-PD)

Cape Wind Off-Shore Wind Energy Project

Case Overview	The MIT Stakeholder-Assisted Representation Process	MIT On-Line Survey
MIT-USGS Science Impact Collaborative	Draft Environmental Impact Statement (November 9, 2004)	Stakeholder Inputs in Scoping Hearing

Stakeholder-Assisted Representation Research at MIT

q   At MIT we are looking at ways stakeholders can be more proactively engaged in the representation and scoping of large -scale engineering projects, such as the Cape Wind project. It is our view that the active involvement of stakeholders can help in making more informed decisions that take into account stakeholder needs and knowledge, and go a long way in avoiding costly conflict in the community.

Phase I. MIT Stakeholder Survey on the Cape Wind Project

q   In the first phase of this research, we have contacted more than 50 stakeholder organizations in the Cape Wind project with a balanced representation of opponents, proponents, undecided and neutral stakeholders, to understand their views on the project and its scope. Download the MIT Stakeholder Survey Report

Phase II. Collaborative Stakeholder Process

In the second phase of our study, we are inviting feedback from stakeholders on the representation created in phase I, to refine it further and make sure stakeholder concerns are adequately covered. Based on that representation, we will then look at the different alternatives and get stakeholder inputs on a potential collaborative approach to the project. Stakeholder feedback can be provided in stakeholder meetings we will be setting up at MIT, or through email and phone.

Fig. 7.30.   Snapshot of Cape Wind case study website .

Collaborative Process Workshop for Offshore Wind Energy Siting

Saturday April 2, 2005 **2:00-6:00 p.m.** **MIT Stata Center, Room 32-124** **32 Vassar Street** **Cambridge, MA 02139** **Campus Map** **RSVP by March 28, 2005 to** **capewind@mit.edu**	• Workshop Details (**PDF**) • Press Release (**March 19, 2005**) • Presentation (**PowerPoint**) • Stakeholder-Assisted Systems Representation (**Vensim Model**)* • **Free** Vensim PLE Software **Download** • MIT Cape Wind Phase I Stakeholder Survey Results

*Steps for Accessing the Systems Representation

(The systems representation is presented in the Power Point Presentation. Follow these steps only if you wish to edit/modify the representation)

1. Download Vensim PLE for free
2. Install on your computer
3. Download the Systems Representation and save in the same directory as Vensim PLE
4. Run Vensim PLE
5. Go to File > Open Model > Cape Wind

Fig. 7.31.   Snapshot of the Cape Wind case study website (workshop announcement).

More than 60 stakeholder organizations for the Cape Wind project were invited and 18 stakeholders, mainly from Cape Cod, agreed to attend the meeting at the MIT Stata Center, a four-hour round-trip journey on a rainy Saturday afternoon (April 2, 2005). Stakeholders who could not attend were given the opportunity to comment via the website or through email. The presentation for the workshop was made available one week before, for the public to be informed of its content. Five stakeholders who were unable to attend the workshop provided feedback on the system representation via the website. The invitation letter is also available in Appendix B.

A press release was circulated among local media outlets, explaining the goal of the workshop (see Fig. 7.32). *MIT Tech Talk* featured the workshop in its March 30, 2005, issue (see Fig. 7.33).

7.6.2   *Participating Stakeholder Organizations*

Stakeholders who participated in the workshop or provided feedback on the representation are listed in Table 7.14.

7.6.3   *Workshop Materials*

Stakeholders were given a folder that included an introduction to joint fact-finding and systems representation, a SAM-PD process diagram (Fig. 7.1), the press release, information about the MIT-USGS Science Impact Collaborative and the initial system representation. Table 7.15 shows a summary of the arguments presented to stakeholders at the workshop and in the handouts.

7.6.4   *Workshop Structure*

The workshop was divided into five parts. In the first part, participants were introduced to the idea of joint fact-finding and collaborative systems representation. In the second part, stakeholders were asked to form four breakout groups to collaboratively refine the system representations for the energy, environment, socioeconomic and navigation, aviation and safety subsystems. In the third part, stakeholder breakout groups were asked to identify four to five major uncertainties in their subsystems that needed to be addressed. In the fourth part of the workshop, stakeholders were asked to identify

BETA

MIT Research Group Explores Stakeholder Involvement in Siting Offshore Wind Energy Projects

The MIT-USGS Science Impact Collaborative (MUSIC) looks at the potential role of joint fact finding as a way to bring together federal and state government permitting agencies, federal, state and local advisory agencies, citizen groups, developer, labor unions, environmental groups, consultants and academic experts in decision-making for offshore wind energy projects.

Cambridge, MA (PRWEB) March 19, 2005 -- The MIT-USGS Science Impact Collaborative (MUSIC) is sponsoring a stakeholder process on offshore wind energy projects at MIT on April 2nd, 2005. Stakeholders in the Cape Wind offshore wind energy proposal have been invited to MIT, to discuss lessons from the ongoing permitting process for the Cape Wind proposal, and look at ways to improve the decision-making process. The stakeholder meeting, to be held in the world renowned MIT Stata Center, will shed light on how to design a siting process for offshore wind energy that takes into account stakeholder concerns and local knowledge, while avoiding competing scientific studies that can undermine the process. "Most stakeholders involved in environmental disputes have different levels of scientific understanding, and trust among stakeholders can erode if each group brings its own scientific resources to support its position, leading to dueling or competing studies and experts. Perceptions of unequal distribution of scientific resources can undermine the collaborative spirit and lead to a breakdown of the process, or worse yet, litigation," according to Dr. Herman Karl, the Co-Director of the MUSIC program. The MIT-USGS Science Impact Collaborative has developed an effective framework called joint fact finding (JFF) to address conflict in complex science-intensive decisions. The purpose of a JFF process is to handle complex scientific and technical questions. JFF helps participants agree on the information they need to collect and how gaps or disagreements among technical sources will be handled. JFF allows stakeholders to build a shared understanding of technical and scientific issues and their implications for policy. They can also help resolve disputes about scientific and technical methods, data, findings and interpretations. The Cape Wind Project is a proposal to build 130 wind turbines off Cape Cod, in Nantucket Sound. The permit process is administered by the U.S. Army Corps of Engineers. With its technical and social complexity, and high degrees of controversy and conflict, the Cape Wind project has become an important case study in the MIT research. The case study was initiated at MIT in 2003 during graduate seminars on joint fact-finding at MIT, where different stakeholders were invited to express their views on the permitting process. In 2004, Ali Mostashari, a Ph.D. candidate at MIT, initiated a research project on stakeholder-assisted modeling and policy design for the Cape Wind project, as the basis of a joint-fact finding process in offshore wind energy siting. As part of this project, more than 44 representative stakeholder organizations were contacted to fill out a survey on their views of the important aspects of the process. Participants in the survey included opponents, proponents, undecided and neutral stakeholders as well as those not taking any positions on the project. Based on the responses to the survey, comments in six public hearings, comments in the Massachusetts Technology Collaborative stakeholder process, stakeholder news releases and interviews, the group was able to construct a system representation that highlighted the most important concerns and issues dealing with offshore wind projects, as perceived by a broad range of stakeholders." Public hearings are just not sufficient to address the complex issues of projects such as Cape Wind. There needs to be a direct stakeholder dialogue, which can bring all sides of the issue together to work out the issues in a collaborative manner. We have seen how conflict can prolong the permitting process, without improving its substance," says Mostashari. "The goal of the April 2nd workshop is to learn from stakeholders in the current permit process how it can be improved for future offshore wind projects, as well as similar issues such as siting of LNG facilities."

Fig. 7.32.   Press release for the collaborative process workshop at MIT.

Volume 49 – Number 22
Wednesday – March 30, 2005

Cape Wind Project meeting

The Cape Wind Project, with its controversial environmental, economic and aesthetic impact on Nantucket Sound and Cape Cod, has been an important case study for participants in MUSIC and proponents of joint fact-finding since 2003. After the issuing of the draft environmental impact statement (DEIS) in November 2004, the permitting agency has yet to announce any decision on the project. The delay is widely seen as politically motivated.

On April 2, MUSIC will host stakeholders in the Cape Wind offshore wind energy proposal to discuss the permitting process and possible ways of resolving some of the scientific issues in dispute. MUSIC interns have prepared a web page (http://scienceimpact.mit.edu) to portray how citizens might use joint fact-finding.

Ali Mostashari, a doctoral candidate in engineering systems, has organized the April 2 meeting. In 2004, as part of his dissertation research project on policy design in the Cape Wind Project, Mostashari convinced more than 44 representative stakeholder organizations to summarize their views of the offshore wind energy siting and permitting process.

"Pubic hearings are just not sufficient to address the complex projects such as Cape Wind. We have seen how conflict can prolong the permitting process, without improving its substance. The goal of the April 2nd workshop is to learn from stakeholders in the current permit process how it can be improved for future offshore wind projects, as well as similar issues such as siting of LNG facilities," Mostashari said.

Fig. 7.33. *MIT Tech Talk* article on stakeholder workshop.

ideal working groups, whose joint findings would be considered credible by all sides. In the final part, stakeholder breakout groups reported back to the entire group with their recommendations on the system representation, uncertainties and working groups. Breakout groups were formed in a way to include a balance of opponents, proponents, undecided and neutral stakeholders. In order to shift the discussion away from the usual discourse, stakeholders were asked to

Table 7.14.   Participating and Contributing Stakeholders in MIT Stakeholder Workshop

Actor group	Organization	Position of representative
U.S. Federal agencies	U.S. Fish and Wildlife Service	Regional director
	U.S. Fish and Wildlife Service	National energy coordinator
	U.S. Geological Survey	Senior technical expert
State and regional Agencies	Massachusetts Division of Marine Fisheries	Cape Wind project contact
	Massachusetts Office of Environmental Protection	Wind energy coordinator
	Cape Light Compact (Regional Consumer Utility Protection Agency)	Cape wind project contact
Local Groups	Alliance to Protect Nantucket Sound	Communications director
	Clean Power Now	President
	Cape Clean Air	President
	Wind, Energy and Ecology Information Services	President
Developer	Cape Wind Associates	Communications director
National Groups	The Humane Society of the U.S.	Field director for marine life
Academia	MIT Ocean Engineering Department	Professor
	MIT Ocean Engineering Department	Graduate student
	MIT Edgerton Center	Lecturer and students
	MIT Engineering Systems Division	Professor
	MIT Department of Urban Studies and Planning	Graduate students
	University of Massachusetts Amherst, Department of Environmental, Earth and Ocean Sciences	Professor

Table 7.15.   Summary of Arguments Presented to Stakeholders during Introduction

Goal of workshop	• Use stakeholders' experience with Cape Wind process to collaboratively design a system representation as a basis for collaborative decision-making processes for offshore wind energy projects
Problems with current permitting process	• Public hearings do not address stakeholder needs adequately • "Everyone is entitled to their own opinions, but not their own facts". Is that so? • Uncertainties in scientific information result in adversarial science, with experts providing different and often conflicting information • Stakeholders spend funds for competing research • Information production for EIS is often perceived as biased by marginalized stakeholders • Transparency is a major problem • Major cost of conflict, litigation, divisions in community • Decisions are often made in courts, taking them out of the hands of stakeholders
Introduction to joint fact-finding	• Joint fact-finding (JFF) is a collaborative process, where federal, state and local decision-makers, citizen groups, private sector, experts and technical experts (in general stakeholders) jointly review a project and negotiate a decision in a consensus-seeking process • Stakeholders focus on underlying interests rather than positions on particular issues. • Current NEPA process can make use of a JFF to inform its decision-making. Section 101 of the NEPA process allows for collaborative decision-making (guidance memos currently circulated in many agencies)

(Continued)

Table 7.15. (*Continued*)

- Joint fact-finding is normally convened by the permitting agency
- If the stakeholder group comes to an agreement, the permitting agency then bases its decision on its recommendations
- Stakeholder-assisted modeling and policy design process is a joint fact-finding process that uses stakeholder-assisted systems representation as a basis for collaborative decision-making
- This process can channel conflict into a productive effort for making the best decision. It does NOT eliminate conflict
- It is no silver bullet, and there is no guarantee that it can stop litigation, but it is definitely more likely to succeed in reducing conflict than the traditional process

---

think in terms of a generic offshore wind energy system representation, rather than the Cape Wind case.

7.6.5 *Ground Rules for Collaborative Representation*

MIT students were assigned to each breakout group as consensus recorders. Modifications and suggestions were only reported as group recommendations when all four or five breakout members came to a consensus to have it included. Each group was given a large poster-size system representation that could be modified and written on. All modifications and suggestions however were recorded separately, regardless of whether agreement was reached. Stakeholders were asked to prioritize three to five modifications and up to five uncertainties per subsystem. They were given a total of 90 minutes for the three tasks (representation, uncertainty identification, working

Fig. 7.34. Initial introduction on joint fact-finding and systems representation.

group proposal). Figures 7.34 to 7.36 show pictures of the MIT Stata Center workshop and the participating stakeholders.

7.7 Workshop Dynamics and Results

In this section we will look at the attitudes, dynamics and behavior of the stakeholders during the introduction, collaborative systems representation, collaborative uncertainty identification and collaborative working group formation.

7.7.1 Stakeholder Responses to the Introduction to Joint Fact-Finding and Systems Representation

In general, while emphasis was made repeatedly that the workshop goal was to go beyond the Cape Wind project, and

Fig. 7.35. Stakeholders working on the energy subsystem representation.

Fig. 7.36. The socioeconomic impact group.

to think whether a joint fact-finding project could be applied to a new offshore wind project, stakeholders kept coming back to their positions on Cape Wind, or ask questions that were mostly implicit statements about their opponents. General statements made during the introduction (para-phrased) are shown in Table 7.16.

Tensions in the session: When a the question was asked by one of the proponents, who went into one of the details of the Cape Wind project, opponents started to argue about the validity of the question. We intervened, by emphasizing that the Cape Wind project was not the emphasis of the session. The intervention however may have made stakeholders feel that they did not get a chance to speak their mind. Tensions remained in the audience until the break, after which the collaborative systems representation was to start. After the introductions were over, stakeholders were offered refreshments and a chance for informal conversations for 15 minutes. There was extensive mingling, and by the end of the break much of the tension seemed to have reduced in intensity.

7.7.2   *Stakeholder Responses to the Systems Representation Breakout Session*

Stakeholders were allowed to self-select into subsystem break-out groups of interest, with some limitations that would ensure balance in the groups. Each group was given a spot in the large room to discuss their respective subsystem representations. Conflict was minimal in the breakout groups, and were limited to grunts when consensus was not reached for particular issues a participant was putting forward. Table 7.17 shows the representation changes made by groups through consensus decision-making. The atmosphere of the groups was mostly friendly, with the exception of one group (socioeconomic

Table 7.16. Comments Made During the Introduction to Joint Fact-Finding

Focus	Questions/comments	Responses given by facilitator/ organizers
Science, facts and uncertainty	— Why is adversarial science considered undesirable? With adversarial science one can get to the bottom of the facts and prove the facts of a process. — There is no uncertainty in the fact that many wind farms in Europe haven't had a single bird kill in the past five years. When people don't accept this, how can this process help?	— Adversarial science does not refer to differing scientific views, but to the competing investments in science that can serve as a weapon in a public dispute. Shared scientific resources and inclusive working groups will allow technical experts representing different stakeholders to examine underlying assumption and methodologies used. — How "historical facts" are interpreted to apply to the current context is a subjective matter. Joint fact-finding allows for a joint clarification of implicit subjective assumptions in making use of knowledge.
Stakeholder rationality	— This process assumes that all stakeholders are rational; we know that there are people here who don't care about the truth or facts.	— Universal rationality is not a basic assumption. However, there is the assumption that people have underlaying interests that determine their positions. If those interests can be considered, positions can often change based on a commonly developed understanding.

(Continued)

Table 7.16.   (*Continued*)

Focus	Questions/comments	Responses given by facilitator/organizers
Process improvement	— How would this process guarantee that political actors wouldn't independently play their politics, regardless of what a group of stakeholders decide? — Who pays for the increased scope of issues that have to be studied? — Stakeholder involvement is difficult when there are limited resources and time constraints in decision-making. — Overwhelming majority decisions are not useful, since it will still leave someone marginalized who can litigate. — This process will take too long. It is the best way for the opponents to stall the project.	— There is no guarantee. What this process accomplishes is to provide a common basis for decision-making among stakeholders. This does not necessarily guarantee success, but it provides an improvement over the dominance of adveraries. — Not all of the representation needs to be included in the scope of the study. That's up to the stakeholder group to negotiate. Who pays is also not pre-determined. Many stakeholder organizations may have resources that could help in this regard. — Resources and time are often wasted in trying to create a litigation-proof EIS, in litigation processes, and in the politicization of the decision-making

(*Continued*)

Table 7.16. (*Continued*)

Focus	Questions/comments	Responses given by facilitator/organizers
		process. Joint fact-finding may take longer to complete on the front end, but may save substantial time for the overall process. For example, the Cape Wind project is now around 2.5–3 years behind schedule. During this time many of the longer contentious studies would have been completed.
Systems representation	— There are issues that a representation cannot capture. For instance how do you capture the fact that the jurisdiction of the project is problematic, and that we lack an effective ocean policy?	— Actually these can be captured by a system representation, as they have been to some extent in the stakeholder process layers and the long-term impact layers of the system representation. Not all parts of the system representation will be quantified into a model, but the effect of institutional issues can be discussed and evaluated qualitatively.

Table 7.17.   Consensus Modifications/Additions to the System Representation by Stakeholders

Breakout group	Layer	Suggested modification/ addition
Energy	Windfarm characteristics	— Wave heights (location)
		— Sea bed conditions
	Energy production	— Long-term (10–20 year) contracts for power
	Energy economics	— Electricity cost stability
		— Financing costs
Environmental impact	Avian and marine impacts	— Seasonal residences more important than proximity to migration routes
		— Construction noise impact impact (on marine mammals) and commercial/recreational fish species
		— Construction impact on benthos. Impact on benthos has direct impact on prey availability and habitat use
	Protected species impact	— Seasonal residence
		— Impact on foraging resources
		— Proximity to migration route of whales
	Air pollution impact	— Number of oil barges and related risks
		— Intake and discharge of water used by plants for cooling, impacting volume and chemicals
		— Reduction in temperature variations
		— Less polluted marine and coastal areas

(Continued)

Table 7.17.   (*Continued*)

Breakout group	Layer	Suggested modification/ addition
Navigation, aviation and safety	Navigation	— Proximity to fishing areas — Flight patterns should also be included — Size of exclusionary zone — Turbines as aid to navigation — Climate conditions affect everything — Risk to search and rescue operations
	Construction and maintenance	— Time of the year restrictions limiting maintenance schedule — Hazmat management — Size of the overall system impacting maintenance costs
Socioeconomic impact	Local and regional economic impact	— Municipal tax revenue for infrastructure > local economic benefit — Revenue recycling > local economic benefit — Transmission system infra-structure impacts local economic costs — Transmission system impro-vements may increase electricity reliability — Source of needed funds limits stakeholder participation — Effect of timing (urgency) limits stakeholder participation — Existence/lack of regula-tions for ocean policy can undermine all positive influences

Note: The refined systems representations are presented in Appendix C.

impact), where it remained cordial throughout. At the end of the first task, most breakout group members seemed to be smiling and having a good time.

7.7.3 *Stakeholder Responses to Uncertainties in the System*

Stakeholders stayed in the same breakout groups, and continued working collaboratively on the most important uncertainties in their subsystems, through consensus. Table 7.18 shows the uncertainties they identified as important.

7.7.4 *Stakeholder Identification of Ideal Working Groups*

Again, staying in the same breakout groups, stakeholders identified working group compositions for their subsystems that would make the technical analysis credible for stakeholders. The focus was a generic offshore wind energy project. Table 7.19 shows the working group formations proposed by stakeholders.

7.7.5 *Reporting Back to the Group*

The consensus recorders within each group brought back the results of the consensus decision-making with regards to the system representation, uncertainty identification and working group formation. The results were then presented to the group at large, which had an opportunity to comment. At the end of the session, stakeholders were asked to fill out a feedback survey on their experience at the workshop, and the potential of joint fact-finding and systems representation. In the following section, we will look at the results of the feedback survey.

Table 7.18.   Consensus Major Uncertainties Identified by Stakeholders

Breakout group	Suggested uncertainties
Energy	• Fluctuations in fossil fuel prices • Cost of renewable energy certificates • Litigation costs • Disaster or terrorism risk
Environment	• How long it would take for power plants to cease operation and power independence to be realized • Delay in air pollution environmental benefits • Comprehensive characterization of resources and habitat use for all endangered and protected species • Modeling the differential avian collision impact • Electromagnetic effects
Navigation	• Human error • Weather • Potential future changes to size of exclusionary zone • Potential future changes in regulatory structure
Socioeconomic impact	• Public acceptance • Monetary health benefits • National energy policy • Cost and benefit distribution • External safety threats (terrorism, etc.)

7.8   Stakeholder Feedback Survey

The survey served as one of the principal ways to explore the validity of the hypothesis of this book. Therefore, the questions were mainly designed to elicit the views of the diverse stakeholder group (including government, private sector, experts and citizen groups) on the criteria for a *superior* presentation.

Table 7.19. Consensus Ideal Working Groups Identified by Stakeholders

Breakout group	Suggested working group composition
Energy	• U.S. DOE • Energy Facilities Siting Board • UMASS Renewable Energy Lab • MIT LFEE • ISO grid manager • Electric Power Research Institute • American Wind Energy Association • Consultants and Experts on Both Sides
Environment	• All federal and state resource agencies • University centers • Local and national NGO representatives • Academic research technical experts • Outreach and communication teams
Navigation	• NOAA • Department of Homeland Security • U.S. Coast Guard • State Coastal Zone Management • U.S. Navy • Shipping line representatives
Socioeconomic Impact	• Chambers of Commerce • Private consultants on all sides • MTC (or equivalent) • State Economic Development Office • State Energy Office • Academia • Congress and state legislature • Fishermen associations

7.8.1   *Inclusion of a Plurality of Views*

The system representation was the result of inputs of tens of stakeholder groups and was therefore by definition inclusive of a plurality of views. We asked stakeholders, however, whether the system representation adequately addressed the key concerns and interests for an offshore wind energy project such as Cape Wind. 67% of respondents said they thought the system representation adequately reflected stakeholder concerns. 25% believed it did not, and 8% believed it would with further refinement.

Here is a summary of comments from people who thought the system representation was either totally inadequate, or needed further refinement.

- Added dimension afforded by seeking to understand the certainty sought by stakeholders would help better refine studies and actions needed to address uncertainties.
- Not yet, but getting there.
- Too light on supportive environmental benefits, benefits to organized labor, and more clear energy benefits.
- The survey was comprehensive, but the meeting was not representative, because there were too few people. Also, some knowledge gets lost when converted into representation.
- I'd like to see more consideration of how this process would change if driven by government, not private sector.

7.8.2 Usefulness of Representation as a Thought Expander for Stakeholders

Stakeholders were also asked whether they felt that their understanding of the offshore wind energy system had improved by working with the system representation. 83% said it had improved their understanding. 8% said that it had helped a little, and another 8% believed it hadn't helped at all. Comments by those whose understanding of the system had only slightly improved or not at all said that they had not had a chance to look at the entire system representation in much detail, or that the issues had been discussed ad nauseum and were known by everyone.

7.8.3 *Usefulness of Representation for Suggesting Strategic Alternatives for Improved Long-term Management of the System*

Stakeholders were also asked whether they thought the system representation would allow for more decision options to be considered than the current permitting process allows. 67% believed it would, 25% believed it may depending on how it is used, and 8% believed that it would not. Stakeholders were also asked whether the representation would form a better basis for scoping offshore wind energy projects than the current permitting process. 83% said that the representation was a better basis for scoping, 8% said that it may be depending on how it is used, and 8% said that it wasn't. Comments by those who were unsure or thought the representation would not support better scoping and/or options are as follows:

- I think it is good to understand relationships of various considerations, but this leaves out personal preferences/values that drive decisions.
- Still needs development.

7.8.4 *Completeness of Representation (Taking into Consideration Technical, Social, Political and Economic Aspects)*

Stakeholders were asked whether the representation was sufficiently comprehensive in capturing the economic, social, political and technical considerations. 75% believed it was sufficiently comprehensive, 17% thought it was not, and 8% said they didn't know.

Those who didn't think it was comprehensive enough had the following comments:

- Needs more refinement, abbreviated language in boxes leads to ambiguity in depicting total impact. The folks who put this together did not know the terms of bibliography well enough to create correct shorthands and draw all the boxes and arrows.
- It's better now, but it still needs work.
- Focus should not be on comprehensiveness but on how to decouple values/opinions from analysis.

7.9 Additional Feedback from the Stakeholder Survey

In addition to issues relating to the hypothesis of this book, we asked stakeholders other questions on the stakeholder-assisted modeling and policy design (SAM-PD) process as a whole.

7.9.1 *Working Group Formation and Science Development through Joint Fact-finding*

Stakeholders were asked whether they thought working group selection by stakeholders in a joint fact-finding context would help reduce conflict and increase the credibility of the analysis. 92% thought it would. 8% were unsure.

7.9.2 *Superiority over Current Permitting Process*

Stakeholders were asked whether they thought the proposed process, based on a commonly agreed system representation, was an improvement over the current permitting process. 83% thought it was and 17% said they were unsure. Comments of one of the stakeholders who were unsure:

- Maybe, only if legislative changes are made to allow the process to impact decision-making.

7.9.3 *Transparency*

We asked stakeholders whether they thought the SAM-PD process would be more transparent than the current permitting process. 75% said they thought it was more transparent, while 25% said they were either unsure, that it may have the potential to be or that it would depend on particular permitting processes.

7.9.4 *Drawbacks of the SAM-PD Process*

Stakeholders were asked to comment on their perceived drawbacks of the SAM-PD process and potential obstacles. Those providing feedback had the following to say:

- Needs to provide a separation of values/opinions from fact-finding/science.
- I don't know that they are drawbacks, so much as areas in need of development. The process needs to be refined and tried on other cases as well.
- Commercially viable sites are very limited. This model would be more effective if there were many options for sites among which one could choose.
- Do it before a project is proposed in order to eliminate sites likely to generate the most controversy and choose sites with the least adverse impact.
- Needs to be institutionalized by permitting agency.

- If it is too narrowly focused it could lead to flawed conclusions.
- There needs to be buy-in by the project proponent.

7.9.5 *Other Stakeholder Comments on the SAM-PD Process*

- It's a new way of looking at an old problem.
- Well, of course this a better process; the traditional requirement for notice and comment make no provisions for stakeholders interacting.
- A rather refreshing look at collaborative, community based decision-making.

7.10 Comparing the Refined Stakeholder-Assisted Representation with the U.S. Army Corps of Engineers Scoping Document

In addition to the stakeholder views, we compared the stakeholder-assisted representation with the scope of the environmental impact assessment used by the Army Corps of Engineers as the basis for decision-making. The scope of the EIS determines the system boundaries, components and interconnections that have to be evaluated for the final decision. There is a major difference between the system representation and the scope of the EIS: the scope of the EIS is essentially a subjectively narrowed down system representation in which the permitting agency has determined what needs to be evaluated. In a SAM-PD process, what remains in the scope is one that stakeholders decide on collaboratively, and within a joint fact-finding context.

Tables 7.20–7.22 show the scope of the EIS. They include components and linkages that need to be addressed for decision-making, and as such can be considered a system representation in a non-diagrammatic format.

Table 7.20.   Scope of the EIS: Avian and Marine Habitat Impact, and Impact on Fisheries

Category	Issues to be studied	Methodology to be used
Avian habitat	— Current use of the final alternative sites by birds as baseline data — Species, number, type of use, and spatial and temporal patterns of use — Issues to be addressed include (1) bird migration, (2) bird flight during storms, foul weather and/or fog conditions, (3) food availability, (4) predation and (5) benthic habitat and benthic food sources — Information derived from other studies, providing a three-year baseline data set — Endangered species impact on piping plover, and roseate tern	— Existing, published and unpublished research results, especially research that describes long-term patterns in use — New field studies undertaken for this EIR/EIS. Data on use throughout the year, especially through November for migratory species, and under a range of conditions. Methods: remote sensing through radar and direct observations through aerial reconnaissance and boat-based surveys — Data gathered through radar to be validated with direct observations — Known impacts to birds from former or current wind turbine generators (WTGs) and other tall, lighted structures (such as communications towers)

(Continued)

Table 7.20. (*Continued*)

Category	Issues to be studied	Methodology to be used
Marine habitat	— Vibration, sound, shading, wave disturbance, alterations to currents and circulation, water quality, scouring, sediment transport, shoreline erosion (landfall) and structural habitat alteration — Northern Atlantic right whale, humpback whale, fin whale, harbor seal and grey seal, loggerhead sea turtle, Kemp's Ridley sea turtle and leatherback sea turtle	— Assessment of (1) species type, life stage, and abundance, based upon existing, publicly available information; (2) potential changes to habitat types and sizes; and (3) the potential for turbines as fish-aggregating structures. The study should assess potential indirect impacts on fish, mammals, and turtles that may result from changes in water movement, sediment transport, and shoreline erosion
Fisheries	— Assessment of potential impacts on specific fishing techniques and gear types used by commercial and recreational fishermen — Multiple-use conflict — The potential for indirect impacts such as changes in fishing techniques, gear type and patterns will need to be included	— Review of existing literature and databases to identify and evaluate commercial and recreational fish data and abundance data in Nantucket Sound — Data to be reviewed should include National Marine Fisheries Service (NMFS) Commercial Data, NMFS Recreational Data, Massachusetts Division of Marine Fisheries Commercial Data, Massachusetts Division of Marine Fisheries Trawl Survey Data and be supplemented with intercept surveys

Source: U.S. Army Corps of Engineers EIS Scope Document.

Table 7.21.   Scope of the EIS: Other Ecosystem and Physical System Impacts

Category	Issues to be studied	Methodology to be used
Benthic	— Sufficient information to compare between alternative marine sites and to provide a general characterization of the benthic habitat of the final sites.	— Assessment and additional data collection as described in the Benthic Sampling and Analysis Protocol
Interactions between benthos, marine and avian food cycles	— Interconnections between the benthic, fisheries and avian resources — Predator-prey interaction data	— Noise and vibration impacts on fish and mammal habitats and migration — Assessment of the magnitude and frequency of underwater noise and vibrations, and the potential for adversely affecting fish and mammal habitats and migration — Assessment of fish and mammal tolerance to noise and vibrations, with particular emphasis on noise and vibration thresholds that may exist for each of the species
Aviation	— Lighting requirements, radar interference and radio frequency interference — Lighting scheme will need to minimize impacts to birds while also providing for safe aviation.	— FAA analysis

(Continued)

Table 7.21. (*Continued*)

Category	Issues to be studied	Methodology to be used
Communication	— Possible impacts to telecommunication systems — microwave transmission	N/A
	— Impact on installation of the wind turbine generators between Martha's Vineyard, Nantucket, and the mainland on existing transmission paths	
	— Impact on boater communications devices	
Navigation	— Commercial and recreational navigation impacts need to be addressed specifically for construction, operation and maintenance and decommissioning	— U.S. Coast Guard risk analysis
	— Cable installation activities to be included	
	— National security issues	
Socioeconomic	— Impacts on electricity rates and reliability in New England	
	— Explanation of any public funding and any applicable tax credits	

(*Continued*)

Table 7.21.   *(Continued)*

Category	Issues to be studied	Methodology to be used
	— Impact on local economy including effects on employment, tourism, boating and fishing, coastal property values and local tax revenues and other fiscal impact to local governments — Environmental justice issues — Educational and tourism impact	
Electric and magnetic fields (EMF)	— Data on potential human health impacts of exposure to 60 Hz EMF and potential impact of EMF produced from wind turbine generators and their associated cables, and the transmission cable	— Identify populations that could be exposed to 60 Hz EMF greater than 85 mG, including humans, fish, marine mammals, and benthic organisms
Air and water pollution	— Compliance with the requirements of the Clean Air Act for construction and operation phases — Potential for impact on the climate of the region — Potential for spills of contaminants into water	— Emergency response plans to mitigate impacts — Construction protocol

Source: U.S. Army Corps of Engineers EIS Scope Document.

Table 7.22. Scope of the EIS: Social Impact

Category	Issues to be studied	Methodology to be used
Aesthetic and landscape/visual assessment	— Visual impacts to any National Register-eligible site in proximity to any of the final alternatives	
Archaeological	— Any impact on historic districts, buildings, sites or objects, local character and culture, tradition, and heritage will be included — Archaeological surveys for final site	— Survey based on previous archaeological and geological investigations — Magnetometer and high-resolution side-scan sonar surveys will be needed to provide electronic data which can be analyzed to assess the potential for any artifacts, such as shipwrecks, followed up by diver reconnaissance where needed — If resources are found which are eligible for listing on the Register of Historic Places, ways to avoid, then minimize, impacts to cultural resources will be considered and discussed. If avoidance is not an option, a Memorandum of Agreement may be required to mitigate potential impacts

(Continued)

Table 7.22. *(Continued)*

Category	Issues to be studied	Methodology to be used
Safety issues	— Safety considerations will include public and employee safety through construction, operation and decommissioning	— Design standards for the structures will be explained. List of preparers will include the names and qualifications of persons who were primarily responsible for preparing the EIS and agency personnel who wrote basic components of the EIS or significant background papers must be identified. The EIS should also list the technical editors who reviewed or edited the statements. Cooperating agencies and their role in the EIS will be listed
Public involvement	— List the dates, locations and nature of all public notices, scoping meetings and hearings. The scoping meeting transcripts and summary of comments report to be provided as an appendix	

Source: U.S. Army Corps of Engineers EIS Scope Document.

7.10.1   *Inclusion of a Plurality of Views*

Having considered all the sources used in eliciting stakeholder input, we can trace the stakeholder-assisted representation back to inputs from more than 70 stakeholder groups and more than 130 individual stakeholders, ranging from federal, state and local government organizations, to environmental groups, experts, local citizen groups and individual Cape Cod residents. The scope of the environmental impact assessment was finalized using the feedback from seven federal government agencies, two state agencies and one regional agency. As such, the stakeholder-assisted representation contains the views of more diverse views, and is more inclusive of a plurality of views.

7.10.2   *Capturing Effects that Expert-only Representation Couldn't Capture*

A fundamental difference between the stakeholder-assisted system representation and the EIS scope is in the scope of decisions each are designed to support. While the scope of the EIS is aimed at producing the knowledge necessary to decide whether or not the permitting agency should approve or deny the permit application for a particular location, or approve it with minor conditions, the stakeholder-assisted system representation is designed such that it can provide a comparison between different alternatives for the long-term management of the system over its entire lifetime. This extends beyond construction, and impacts the design and management of the system based on its emerging behavior.

7.10.2.1   *Energy*

The EIS only looks at the impact of the proposed wind energy plant on local electricity rates and lacks a comprehensive energy subsystem. This is not surprising, since the permitting decision is not focused on the overall impact of the proposed plant on the regional energy issues. The stakeholder-assisted system representation looks in detail at energy consumption, energy production, energy economics and wind farm characteristics in a comprehensive and holistic manner. The differences arise in the definition of the problem and the system boundaries. The stakeholder-assisted system representation enables the exploration of the following questions, which cannot be answered with the EIS scope:

1. How much energy demand will there be in the coming years in the Cape and Islands?
2. How much available energy will there be if no new plants are built?
3. How would investments in different energy capacities impact energy supply?
4. What does it take for 5% of the energy supply in The Cape and Islands to come from renewable sources by 2010?
5. What are the impacts of the different tax incentives, green credits, environmental research funds, mitigation funds, capital interest and fossil-fuel price fluctuations on the rate of electricity in the region?
6. What effect would different leasing policies have on energy economics for offshore wind energy?
7. How competitive would offshore wind energy be, given a range of different oil and gas prices?
8. What is the impact of direct distribution to towns versus grid distribution on electricity costs?
9. What combination of design parameters would make the project economically feasible?
10. What is the impact of moving the project from the proposed place to another one in terms of economic and technical feasibility?

7.10.2.2 *Environmental impact*

The environmental impact subsystem of the EIS is rather comprehensive and comparable to the stakeholder-assisted system representation in terms of the components and linkages it explores. The differences are minor, and revolve around different emphases on which endangered bird species are to be studied. The EIS lacks sea ducks as a component, while the stakeholder-assisted representation lacks piping plovers. What is missing from the scope of the EIS is a comprehensive approach to the global impacts of GHG reduction and the potential impact of air pollution reduction on public health in the region.

7.10.2.3 *Socioeconomic impact*

In terms of socioeconomic impact, the two representations are comparable. The main difference is the existence of a long-term impact layer in the stakeholder-assisted representation, which looks at the project as a precedent-setting case.

7.10.2.4 *Navigation, aviation and safety impacts*

The two representations are again quite similar in terms of navigation and aviation impacts. In addition to the issues covered in the EIS, the stakeholder-assisted representation considers challenges to search and rescue helicopter operations, and the potential for upgrading turbines with more advanced technologies (flexible turbine designs).

7.10.2.5 *Public involvement*

The stakeholder process layer of the stakeholder-assisted system representation is far more comprehensive than its EIS counterpart, capturing the effect of conflict and potential delays on project implementation on the overall cost of the wind energy proposal and tying participation and conflict resolution back into project economics.

7.10.3   Usefulness of Representation in Suggesting Strategic Alternatives for Improved Long-term Management of the System

The stakeholder-assisted system representation allows decision-makers to explore a variety of decision options in addition to the Yes/No decisions on a particular location. Table 7.23 shows the Yes/No alternatives, along with four additional packages that could help address many of the uncertainties of the project, even within the current location.

The packages in Table 7.26 are just a few sample packages that such a system representation could produce. Many more can be created using different combinations of the policy levers.

The main difference with the traditional permitting process is thus the possibility of *contingent* agreements. Contingent agreements allow for project approval contingent on future actions given particular circumstances.

7.10.4   Completeness of Representation (Taking into Consideration Technical, Social, Political and Economic Aspects)

As indicated earlier, the stakeholder-assisted representation addresses most of the issues addressed in the EIS, while also addressing other crucial issues that stakeholders view as part of the problem definition. As the stakeholder input in Section 7.8 indicated, the existence of the extensive regional

Table 7.23. **Sample Alternative Packages Identified Based on Stakeholder-Assisted System Representation**

Package	1	2	3 (Yes option)	4	5	No build
Number of turbines 2006	80	130	130	130	80	
Number of turbines 2010 if impacts as anticipated	130	130	130	130	130	
Continuous monitoring	Yes	Yes	No	Yes	Yes	
Mitigation insurance fund	No	Yes	No	Yes	Yes	
Environmental research fund	Yes	Yes	No	Yes	Yes	
Lease payment	Yes	Yes	Yes	No	Yes	
Shutdown during peak migration season	Yes	Yes	No	Yes	Yes	
Potential removal of particular turbines with disproportionate higher impact	No	Yes	No	Yes	Yes	
Return of mitigation insurance fund to developer if impacts equal or less than anticipated	No	Yes	No	Yes	Yes	
Favorable support to expand Wind farm further if impacts less than anticipated	No	No	No	No	Yes	
Developer required to upgrade with better technology every 10 years	No	No	No	Yes	Yes	

energy subsystem, the long-term impact layer, the stake-holder process layer and the post-construction monitoring layer allows for greater accuracy in addressing the issue at hand.

7.11   Chapter Summary

In this chapter we applied the SAM-PD process to the Cape Wind Offshore Wind Energy Project. Specifically, we engaged stakeholders in the system representation, uncertainty identification and working group formation within a joint fact-finding context. We then compared the resulting system representation to the scope of the environmental impact assessment process and found that the stakeholder-assisted representation was more comprehensive, provided more decision-making options, captured effects that the scope could not capture, and included a plurality of views. Stakeholder survey results during a SAM-PD workshop also confirmed these observations. In the next chapter, we will look at the lessons learned from the Cape Wind case study, including refinements needed in the SAM-PD process based on the actual application of the process.

Chapter 8

Learning from Cape Wind

Everything has been said before, but since nobody listens we have to keep going back and beginning all over again.

— André Gide, Le Traité du Narcisse

The Cape Wind case study outlined in the previous chapters provides interesting insights into some of the merits and drawbacks of collaborative systems representation, and provides an opportunity to refine the SAM-PD process. In this chapter we will reflect on these observations as they apply to different aspects of collaborative processes. To what extent these observations may apply to other collaborative processes is for the reader to decide.

8.1 SAM-PD Process Preparation

Process preparation is one of the most crucial elements for the success of any collaborative process. While every effort was made to overcome shortcomings with advanced planning and persistence, there were many obstacles, which did not allow for the SAM-PD process to be applied seamlessly. The main obstacles revolved around the fact that in order for a SAM-PD process to succeed, it has to be convened by an entity that is a formal part of the decision-making process. For the Cape Wind case study, there was no entry point into the formal decision-making process for the research team. Additional issues included timing, critical stakeholder involvement and insufficient face-to-face access to stakeholders.

8.1.1 *Importance of Involving the Decision-makers as Conveners*

The Cape Wind case study emphasized the importance of having actual decision-makers as process conveners. One of the main disadvantages experienced throughout the case study was that we lacked the authority of a convener-appointed neutral. We indicated earlier that neutrals and facilitators have to be appointed by conveners and the stake-holder group respectively. A self-appointed neutral can only count on the interest and enthusiasm of stakeholders as well as the high degree of controversy in a project to motivate stakeholders to take part in a collaborative process. In the case of Cape Wind, the name of MIT worked well in attracting stakeholders. This was due to a couple of factors. For some stakeholder groups, MIT presented the potential of a heavy-weight ally or opponent for their cause. For other stakeholder groups, MIT's scientific and technical credibility seemed to ensure its support of renewable energy development.[26] Others thought that a stakeholder process would be a good way to delay the project, by undermining the validity of the current NEPA process. Still, we had to rely on extensive indirect stakeholder inputs (such as public hearings, press releases

---

[26] When sending out emails, my signature read "Martin Fellow for Sustainability". On two occasions Cape Wind proponents responding to my emails assumed in their responses that I was a supporter of the wind project. This however did not seem to create any problems with the opponents of the project, since they saw my research as a tool to delay the project and declare the NEPA process bankrupt.

and newspaper articles) for many crucial stakeholders who did not take part in our survey or participate in the stakeholder-assisted representation workshop. For most stakeholders, it took repeated requests to have them respond to the value assessment survey.

This would have been different if SAM-PD had been used by the permitting agency (U.S. Army Corps of Engineers) or the Massachusetts Technology Collaborative (MTC). After participating in the Cape Wind workshop on April 2, 2005, the Cape and Islands Renewable Energy Collaborative (CIREC) started a coordinated community energy planning for the Cape and Islands, and opted to use a SAM-PD process for collaboratively identifying policies at the household, town and regional levels that would lead to a fossil-free Cape Cod by 2025. With the MTC and CIREC spearheading the project, many of the stakeholders we did not have access to filled out the stakeholder value assessment survey and are participating in the collaborative process.

When the permitting agency is part of the convening group for a stakeholder process, there is a much higher perceived chance of the recommendations and agreements affecting the actual decision. This increases the attractiveness of the process for stakeholders, and gives it more legitimacy and formality.

8.1.2 *"Right" Timing for SAM-PD Process*

An important research question that needs to be addressed empirically in the literature is the issue of timing of a collaborative process. The SAM-PD process is designed with the implicit assumption that stakeholders would be involved just after the initial problem definition stage of a decision-making process. Essentially this constitutes the earliest that a collaborative process can be initiated, since issues have to be at least initially defined and stakeholder groups have to be formed and ready to participate. Yet, as we move forward in the decision-making process we face trade-offs. On the one hand, all the crucial stakeholder groups will be easier to identify so that no key stakeholder groups emerge later in the process. Also, a little emerging controversy is healthy since it helps keep stakeholders interested and engaged in the process. On the other hand, collaborative processes take time, and decision-making windows are limited. Besides, if stakeholders are already polarized to the extent that a collaborative process cannot build initial trust among them, it may be too late for the process.

In the Cape Wind case study, the ideal timing for a SAM-PD process would have been during the MTC stakeholder process in late 2002, when the issues were still being explored and the EIS scope had not been finalized. Unfortunately, we had not yet started with our case study at that time. However, it is important to note that other types of collaborative processes may be applied in later stages of a decision-making process. Because of the centrality of a stakeholder-assisted system representation in SAM-PD, its application is more or less limited to earlier stages of the decision-making process.

8.1.3 Critical Stakeholder Involvement

While many of the crucial stakeholders took part in the Cape Wind case study, an important set of stakeholder groups were not motivated to participate. These included the various fishermen associations, the U.S. Army Corps of Engineers and the State Environmental Protection Act Office. The former did not respond to repeated emails and calls, and the latter two excused themselves since they were involved in the decision-making process and did not wish to give the impression that this was a parallel process to the actual permitting process. This again demonstrates the need for an entry point into the actual permitting process, and the lack of legitimacy for a parallel process.

Additionally, the representative of a contacted organization can also have an impact on the dynamics of the process. We observed that the higher-ranking members of organizations had a more open hand in collaborating, since they were less concerned that their statements might be seen as representing their organizations.

8.1.4 *Necessity of Extensive Face-to-Face Interviews in the Value Assessment Stage*

While we elicited inputs from more than 190 stakeholder groups and individuals through online surveys, public hearing statements, press releases and newspaper articles, nothing seemed more effective than personal contact. Those stakeholders with whom we interacted on a more personal level, and face-to-face, were more likely to participate enthusiastically in the process or suggest other stakeholders. Face-to-face interactions allowed stakeholders to build more trust and become more interested on a more personal level in the collaborative process. Therefore, while online surveys can make the inputs more structured and more useful for systems representation purposes, they fail to build trust by themselves.[27] In the Cape Wind case study, the personal touch was a result of personal interviews with a few stakeholder

---

[27] One of the stakeholders told me informally that my Middle Eastern name may have made it harder to convince people in less urban areas of Cape Cod to participate. He then hastily added, "But if they had seen you, they would have liked you a lot. You talk and sound very American to me".

groups, attending public hearings, personally handing out over 300 research pamphlets to stakeholders and chatting informally about the process.

8.2 Collaborative Process Dynamics

The face-to-face portion of the collaborative process in the Cape Wind case study, that is, the stakeholder workshop, was rather brief and did not allow for the observation of many interesting phenomenon that would shed light on its dynamics. Still, the importance of the role of the facilitator emerged as an important consideration in a SAM-PD process. Aside from the skills and expertise of a professional facilitator, it is important for the system model gatekeeper (or modeler) to be a different person from the facilitator. Early in the collaborative process, some stakeholders made efforts at derailing the process by going back to detailed positions at a time when the larger picture was being discussed. In these instances it is imperative to have a facilitator other than the modeler, who can steer the discussions back to the initial process structure. It is important that the facilitator has the skill to avoid the domination of the discussions by the more aggressive stakeholders.

In the case of Cape Wind, the modeler played the role of the facilitator for the first part of the workshop, and various MIT students were briefed on how to facilitate in the breakout sessions and then appointed to different groups. In the first part of the workshop, where the idea of collaborative processes and an overview of the system representation were presented, two stakeholder groups got into a verbal disagreement. The modeler had to cut both of them short abruptly, and emphasize that they could continue the discussion if it pertained to a particular breakout session later on. This did not sit well with the stakeholders who felt they had not been heard.

While breakout facilitators were given instructions and ground rules on how to manage the groups and record agreements and disagreements, their lack of experience led to the more aggressive stakeholders being more dominant in the discussions. The ground rule of consensus-only modifications and suggestions helped in limiting the impact of this dynamic, but one could argue that with more experienced facilitation the breakout dynamic may have been more tractable. Also important was the necessity of contextual knowledge for each breakout group facilitator. We later established with the stakeholders that what had been recorded in one of the breakout groups (Environment) was not what the breakout group had in mind. The facilitator of the breakout group was unfamiliar with some of the technical terms and therefore recorded a few of the ideas incorrectly.

8.3 System Representation as a Basis for Collaborative Process

While the basic structure of collaborative processes is similar, the use of a system representation can change some of the dynamics of the process.

8.3.1 *System Representation as a Knowledge Organization Tool*

What the majority of stakeholders agreed on was that a system representation can be a good way to structure dialogue about a problem. Many stakeholders had experiences with past collaborative processes where extensive laundry lists were created but never put into context. For most stakeholders it was refreshing to see how their views fit into the larger picture of the offshore wind energy system and interacted with other components. One stakeholder commented that it was not unlike putting together a puzzle or doing the *New York Times* crossword puzzles.

Most stakeholders found the term "system representation" to be misleading. They preferred the term "model". Many said they would not distinguish between a qualitative model and a quantitative one, because a qualitative model could later be quantified or mostly quantified. While it may make sense to distinguish between the two in academic settings, for practical purposes it is better to refer to the system representation as a system model.[28]

8.3.2 *System Representation as a Trust-Building Tool*

The Cape Wind project has a history of bitter community relations. Many of the people invited to the workshop resented one another. Most had never talked to each other directly or held a constructive dialogue. Given the nature of the system representation, where the emphasis is mainly on problem definition, rather than risk assessment, cost distribution and decision-making, this stage created an opportunity for stakeholders to interact without the concern of having to compromise on anything. The stakes in a system representation are perceived as rather low, and the stage is far enough from the final decision-making stage that it allows stakeholders to be flexible.

With the time limitations announced, the teams started to work on modifying the system representation. While it was initially awkward for people to interact, the feeling of creating a common product gradually led to interesting rivalries among different breakout groups. Consisting of stakeholders with differing views on Cape Wind, the teams came to consider their subsystem as a common product. Many of the teams would try

---

[28] Most stakeholders didn't even like the term "conceptual model", since that didn't seem adequate for the system representation. The dominant preference was to call it a "system model".

to finish one section ahead of time, asking how the other teams were doing in terms of time, and whether any other group had come to an agreement on the representation. What started out as a skeptical group of people with personal reservations gradually gave way to a sense of enthusiasm.

This seems to suggest that if used from earlier stages of a collaborative process, a system representation can serve as a means for people to work together in different ways than they are used to, accelerating the trust-building phase that is of crucial importance to collaborative processes.

8.3.3 *Ease of Interaction with a System Representation*

The concept of a system representation was intuitive to most stakeholders. It took stakeholders very little time to familiarize themselves with creating system representations. However, some stakeholders had problems in understanding the concept of polarities or the signs on the directional arrows. In a system representation a positive arrow going from one component to another means that an increase in the effect of the first component will lead to an increase in the effect of the target component. In other words, it is an issue of directionality of change rather than a positive or negative influence. While this had been described to stakeholders during the presentations, some stakeholders found this to be non-intuitive. While polarities are crucial to understanding the dynamics of a system, it may be better to introduce them at later stages, when stakeholders are comfortable with the basic concepts of systems representation.

8.3.4 *Importance of Policy Levers*

Stakeholders particularly liked the concept of "policy levers". As described in Chapter 2 policy levers are components of the system that can be "tweaked" to affect the system as a whole.

For most stakeholders the rationale behind a system representation only became clear when they understood that there are components in the system that are decision variables. Most stakeholders felt more in control of the system representation after this point was explained to them. When they started to look at the subsystems, most teams started by looking at the policy levers and working their way through the representation. Stakeholders commented that the existence of so many policy levers would mean a wider range of decisions that could be made.

8.3.5 *System Representation as a Working Team Formation Tool*

The ability of stakeholders to look at different parts of the system as a whole rather than at individual issues in a laundry list provides an opportunity to assign different working groups to evaluate different parts of the system. In the workshop, stakeholders initially defined ideal working groups that could be assigned to various aspects of offshore wind energy projects. Many had suggestions on how to lump different linkages into one working group, due to the similarity of expertise and resources needed. Overall, stakeholders found that having a system representation would allow them to make sure that all the important aspects of the problem were covered and could be assigned to different working groups.

8.3.6 *Systems Representation and Uncertainty*

Prioritizing uncertainty was one of the first more controversial issues in groups. Stakeholders realized that uncertain areas were more prone to be included in the evaluation and assessment stage. Opponents therefore emphasized uncertainties that were hard to reduce (environmental), while proponents focused on uncertainties of regulation, market, terrorism, etc. The prioritized list shown in Table 7.18 essentially was a compromise between the different stakeholders. The system representation allowed stakeholders to see how those uncertainties would impact the system as a whole, rather than a particular component. This allowed discussions in the breakout groups on whether an uncertainty, however large, would have an important impact on the system. The idea of color-coding uncertain links was met with mixed reactions. Proponents did not like red as the color for large uncertainties and believed such a color-coding could give the perception that the project was flawed. Opponents were rather fond of the idea. In hindsight, showing uncertainty in the system representation has to be rethought and refined.

8.3.7 *Need for Quantification*

Nearly all stakeholders assumed that the current system representation would have to be quantified in order to be useful. Parts of the representation dealing with institutional issues could be left as contextual and qualitative considerations, but stakeholders seemed to see one of the advantages of the system representation in the Vensim environment to be the possibility of quantification. What seemed attractive to stakeholders was the ability to look at dozens of alternative strategies and potentially compare their impacts across the different performance metrics identified. For this reason, stakeholders see the system representation and its subsequent quantification as a promising tool throughout the decision-making process.

8.4   Compatibility of SAM-PD with Current Permitting Process

Most stakeholders were vague on how a SAM-PD process might fit into the current permitting process. Some had initially assumed that the idea was to have system representations to capture ideas voiced during the public hearings. Others saw this process as a promising parallel process that would produce recommendations the permitting agency could then use as an additional criterion for its decisions. The general impression was that without serious changes to the current NEPA process such a process would have limited value. On the other hand, stakeholders agreed that it would be necessary to show the merits of such a process in action before any steps could be taken to make it part of the NEPA process.

In the case of Cape Wind, most stakeholders thought that this process came too late in the process. They would have preferred to see such the process in late 2002 instead. The developer was skeptical about any role of collaborative processes in the current permitting process for the Cape Wind project, and had therefore requested that the workshop focus on post-Cape Wind offshore wind energy projects.

Jennifer Peyser,[29] an MIT graduate student working with the MIT-USGS Science Impact Collaborative (MUSIC), has looked at the potential of incorporating collaborative processes into the current NEPA process for offshore wind. She argues that Section 101 of the National Environmental Protection Act, which focuses on "productive harmony" between man and nature, can be interpreted as an imperative for collaborative processes.

Essentially, collaborative processes could enter at any stage of the NEPA process in different forms. However, a system-representation-centered collaborative process is more suited to the initial stages of the decision-making process and can currently serve as a parallel process with actual NEPA processes, providing consensus-based recommendations,

---

[29] Jennifer Peyser, "Joint fact finding for public involvement in wind-permitting decisions: Beyond NEPA", MUSIC Website, http://web.mit.edu/dusp/epg/music/pdf/peyser_000.pdf (last accessed April 15, 2005).

which the permitting agencies would most probably consider an important basis for their decisions.

8.5   Conclusion

In the process of writing this book and going through the case studies, some observations on stakeholder involvement and system representation emerged as important. A summary of these observations, which often reinforce existing knowledge but also offer new findings, is presented.

8.5.1   *Stakeholder Involvement*

Importance of strong convener and compatible legal framework: For any collaborative process to be successful, the convening organizations need to have the power, legitimacy and resources to push the process through and have the ability to enforce its agreements, overcoming social, economic and political obstacles that emerge on the way. If the genuine will and resources for the process do not exist within the convening agencies, there is little chance for the process to succeed, regardless of its design and management. Therefore, unless the organizational mandate of the conveners and the legal framework is compatible with a collaborative process, the benefits of a collaborative process will remain limited.

Importance of facilitator: SAM-PD tries to limit obstructionism partially by starting with an initial system representation and insisting on ground rules for the collaborative process. However, thoughtful process design is a necessary but insufficient condition for the success of collaborative process. Like any other collaborative process, the success or failure of the SAM-PD process depends significantly on the skills of the facilitator in managing stakeholder interactions and channeling conflict into productive cooperation on the process.

This subjective dependence on the facilitator may be considered a drawback of the process from an engineering perspective, but the importance of 'skill' and experience in the success or failure of a project is by no means new to the engineering profession.

Choice of stakeholders: For any large-scale engineering project there are often tens of thousands of stakeholders and dozens of major stakeholder groups. Not all of these stakeholders can be involved, and not all are interested in being involved. The theoretical level of participation can range from minimal (in an autocracy) to all-inclusive (in a direct democracy). The actual level of participation will be somewhere between the two (see Figure 8.1). Additionally, the value of increased stakeholder participation does not increase uniformly. Intuitively, there will be a place where the incremental value of additional stakeholder involvement will have leveled off or even have decreasing benefits by increasing the cost and complexity of the process to an intractable

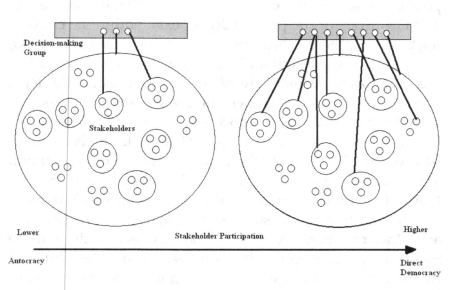

Fig. 8.1.   Increasing stakeholder participation in decision-making.

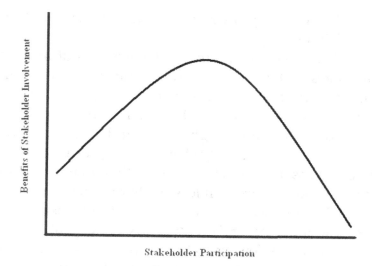

Fig. 8.2.   Incremental benefit of stakeholder participation.

level. This idea is captured in the hypothetical benefit-participation curve in Figure 8.2.

Therefore, it is imperative that the "right" stakeholders be at the table in a collaborative process. The SPK framework was proposed to guide the thought process of the neutral in mapping key stakeholders. In essence the idea behind the SPK framework is that the stakeholder group that participates in the collaborative process must add value to the process, by increasing its collective enforcement power, legitimacy, resources, knowledge or expertise. However, this is a crucial area where there is much potential for more empirical research.

In addition to which stakeholder entities should be involved, the choice of the particular individual representing his organization or a group of stakeholders makes a difference in the process dynamics. Usually, higher-ranking members of an organization have a more open hand in collaborating, since they are confident that their statements may be seen as representing their organizations, and have an easier task convincing their own organizations of a decision in the collaborative process.

Lifecycle cost of a collaborative engineering system decision-making process: One of the main obstacles cited for using collaborative processes for decision-making is the extra cost associated with involving stakeholders. One could argue, however, that the initial higher cost and increased process time which occur in the early stages of the process can reduce the time and cost of conflict and delays that occur with expert-based processes, and make it less costly to deal with emergent adverse behaviors of a system. As an illustrative example, one could consider the Super 7 highway in Connecticut, where the expert process took four months, but the project was delayed for 17 years and was only partially completed due to strong stakeholder resistance. This idea is captured in Figure 8.3.

8.5.2 *System Representation*

System representation as a knowledge mapping tool: A system representation can be a good way to structure dialogue on a problem. For most stakeholders, seeing how their interests in parts of the system fit into the larger picture results in an improved understanding of the system. Also, a system representation can identify where most of the knowledge is concentrated, and which areas still lack information.

System representation as an integrated assessment tool: Most of this book has focused on how various stakeholders can jointly work on a complex engineering system through a common system representation. However, a system representation can also serve as an integrated assessment tool for a complex system among a group of experts coming from different disciplines and focusing on different parts of the system. The Mexico City project is a prime example of this necessity. With different expert subgroups each going off on his own tangent,

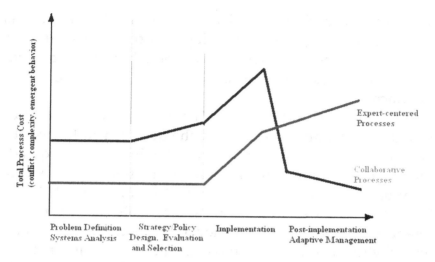

Fig. 8.3. Total process cost for collaborative and expert-centered processes.

with little communication and varying assumptions, the final integration of results became a complicated and nearly impossible task. A common system representation can help communication among experts, even in cases where public involvement is not present.

System representation as a trust-building tool: A system representation can serve as a way for people to work together in different ways than they are used to, accelerating the trust-building phase that is of crucial importance to collaborative processes. This trust is badly needed as a momentum for the rest of the collaborative process, where value conflicts emerge more strongly.

Ease of interaction with a system representation: The concept of a system representation that shows issues and their interconnections in a semi-graphical form is intuitive to most stakeholders. Designing more interactive and graphical interfaces can increase the ease with which stakeholders interact with the system analysis. This can include features that allow stakeholders to click on links to see available sets of data and trace the progress of working groups in analyzing the links. There is, however, the consideration that "technophobic" stakeholders may be more at ease with less technical interfaces.

Importance of policy levers: The identification of policy levers, that is, areas where stakeholders can influence the system, is very empowering for stakeholders. It both expands their decision space and focuses their attention on which components in the system are actually important. This can then serve as a way to generate strategic alternatives by different combinations of policy levers. Essentially, for x policy levers with y choices each, there are a total of x^y alternatives. This creates much flexibility in the negotiation stage.

System representation as a working team formation tool: The ability of stakeholders to look at different parts of the system as a whole rather than at individual issues in a laundry list presents an opportunity to assign different working groups to evaluate different parts of the system. Stakeholders can form groups that include experts they trust to represent their views, and who can defend their interests in the knowledge generation process. The system representation can then show which parts of the system have been covered by existing groups, and any additional experts who may be needed to address other parts of the system the stakeholders agree to study.

Systems representation and uncertainty: Prioritizing uncertainty is one of the more controversial issues in groups. Stakeholders realize that uncertain areas are more prone to be included in the evaluation and assessment stage. The prioritized uncertainties are usually a compromise between different stakeholders. The system representation allows stakeholders to see how those uncertainties would impact the system as a whole, rather than a particular component. This allows discussions in the breakout groups on whether an uncertainty, however large, would have an important impact on the system. Even if no information exists on parts of the system, expert estimates can provide an upper bound for uncertainties.

Quantification and evaluation of system representation: By not specifying a particular model for quantifying or evaluating different relationships in the system representation, SAM-PD allows the integration of many different models into an integrated system model. By linking the different inputs and outputs of various subsystem models, SAM-PD can help evaluate the impact of different alternatives on the over-all system. Another insight was that not all of the system representation need, to be or can be quantified. Social and institutional components and interconnections can be evaluated with social science frameworks, and many quantifiable components with a lack of baseline or predictive data may be considered in the decision-making, but not quantified. Still, having these components allows us to monitor and measure them at later times, or understand an emerging impact on the system.

One of the prerequisites for SAM-PD or other collaborative processes is the existence of some institutional support and legal framework compatibility for collaborative processes. While this may be present in varying degrees in the context of more developed countries, it can certainly pose a challenge in developing countries with lower institutional capacity and a shortage of expertise and resources.

Additional challenges are the potential lack of organized stakeholder groups that are representative of the stakeholders at large. In some situations, special interest groups may be more organized than the less vocal components of the population, leading the collaborative process towards a narrow outcome that may benefit few stakeholders. Additionally, there are cases where stakeholders may not want to participate in a collaborative process, intent on derailing a policy decision regardless of the outcome. Of course these problems can also occur in developed countries, although to a lesser extent.

More challenges can arise when a project is overly politicized, making a "wise" decision irrelevant. In such cases it may be impossible to shift the decision-making process from an ideological and political battlefield to a collaborative wisdom-seeking process. The national missile defense system may be a good example of such systems.

The central idea of this book is that current stakeholder involvement approaches for large-scale engineering systems are inadequate, and that *effective* stakeholder involvement in engineering systems decision-making is essential to the process. The emphasis on "effective" refers to the fact that not all stakeholder involvement results in improved representation, design and management of engineering systems.

This book proposes the stakeholder-assisted modeling and policy design (SAM-PD) process as an effective collaborative process for engineering systems decision-making that allows stakeholder involvement from the problem definition stage to implementation and post-implementation management of a project. The SAM-PD process is a synthesis of existing collaborative processes in the negotiation and conflict resolution literature and existing engineering systems analysis processes. Through the application of the process to three actual engineering systems projects, the book showed that stakeholder involvement in engineering systems design and management did not undermine the technical quality of the analysis, and in fact added to the quality of the system representation that serves as a basis for the engineering systems analysis and decision-making.

As such it emphasizes evaluative complexity as a crucial aspect of engineering systems and concludes that systems analysis cannot be separated from the decision-making and implementation for engineering systems in their complex social and institutional environment.

The basic implication of this research is that it would be myopic of engineering systems analysts to shift the burden of stakeholder involvement to decision-makers, and keep the analysis merely an expert-centered process. Due to the many subjective choices that have to be made with regards to system boundaries, choice of components, inclusion of linkages, nature of outputs and performance metrics and assumptions about data and relationships, systems analysts are in fact not producing the analysis that will help the decision-making process. The best airport designs done with multi-trade-off analysis and intricate options analysis may lead nowhere if stakeholders affected by the project do not see their interests reflected in the analysis. The idea is that a good systems analysis is not one that impresses other engineering systems professionals with its complexity, but one that can actually address the problems at hand.

SAM-PD provides an alternative engineering systems analysis process, whereby the combined wisdom of decision-makers, experts, citizen groups and private sector actors determines the design space, system boundaries and model outputs that are useful for decision-making.

Without a change in the way engineering systems analysis is done, many more projects will have the same fate as the Cape Wind Offshore Wind Energy Project, the Mexico City Airport, the Yucca Mountain nuclear waste repository, and the Connecticut Super 7 highway, which had brilliant technical designs but failed to take into consideration that the aim of building engineering systems is to serve the actual needs of people, not just to create another technical artifact with a high price tag and impressive technology.

Index

Actor groups   44, 172–175, 215
Actors   4, 18, 19, 112, 117, 136,
    254, 299
Adaptive Management   9, 105,
    112, 153, 185, 187
Advocacy   56, 61, 72
Alternative (Strategic Alternative)
    18, 44, 184, 262, 276, 296

BATNA   145
Bias   56, 57, 60, 74, 83, 92
Bounding   57

Cape Wind   1–3, 81, 154,
    157–160, 161, 170–173, 175–177,
    179–181, 183–192, 194–197, 202,
    205–209, 213–217, 223, 225, 236,
    241–243, 247–250, 252, 255, 261,
    278–284, 286, 290, 299
Checklist   44
CLIOS Process   13, 32, 42–45, 47,
    48, 86, 88, 106, 107, 127, 133,
    136, 154, 214
CLIOS Systems   85, 215
Complex System   13, 15, 17,
    24–26, 27, 31–33, 35, 40, 42, 46,
    51, 52, 153, 294
Complexity   7, 9, 12, 18–21,
    24–28, 31, 33, 39, 43, 45, 46, 54,

55, 59, 83, 96, 97, 152, 215, 245,
    292, 298, 299
Compromise   286, 289, 296
Concessions   145
Conflict Resolution   13, 114, 121,
    276, 298
Consensus-building   106–108,
    121, 146
Constituents/Constituency   5,
    146

Discourse Integration   107, 108,
    122–126
Drivers   91, 94, 123, 126, 133, 134,
    176, 219, 223, 225
Dynamic Complexity   19
Dynamics   13–16, 18, 32, 36–43,
    45–48, 52, 69, 86, 87, 120, 141,
    143, 145, 250, 283–285, 287, 293

Emergence   9, 18, 20, 24, 31
Engineering Design   10, 20
Engineering Systems   1, 4–10, 13,
    16–21, 23, 26, 27, 31, 32, 34, 35,
    41–43, 45–48, 51–54, 56, 57, 65,
    66, 72, 77, 83, 85, 86, 89, 95, 96,
    98, 102, 105–107, 109–111, 113,
    114, 128, 147, 151, 153, 154, 247,
    298, 299

Environmental Impact Statement
2, 158, 159, 161, 162, 171, 177,
213, 242
Evaluative Complexity 7, 19, 20,
25, 33, 45, 97, 298
Expert analysis 7, 51, 99, 150,
184

Facilitation 100, 101, 118, 121,
285
Facilitator 75, 99, 114, 121, 129,
132, 137, 138, 140, 145, 146, 177,
185, 253–255, 280, 284, 285, 291,
292
Framework 13, 32, 44, 105–108,
116, 117, 144, 220, 245, 291, 293,
297
Framing 11, 55, 67, 102, 129

Game Theory 32, 35–37
General Systems Theory (GST)
23
Genetic Algorithm (GA) 32, 34

Impartiality 121
Institutional Sphere 43, 44, 122,
133, 215
Interest groups 4, 297
Interests 7–9, 11, 59, 67, 102, 112,
117, 118, 121, 126, 129, 131, 138,
139, 189, 203, 207, 248, 253, 261,
294, 296, 299
Internal Complexity 18

Joint Fact-Finding 16, 121,
127, 130, 131, 137, 138,
142–144, 177, 244, 245,
248–250, 252, 253, 255, 258,
263, 265, 278

Knowledge 9, 14, 16, 27, 38, 51,
52, 54, 56, 58–65, 67–70, 73, 74,
78, 79, 83, 85, 88, 89, 96, 97,
106–108, 113–116, 119, 122,
125, 128, 131, 133, 136, 137,
139, 144, 153, 154, 189–192,
242, 245, 253, 261, 273, 285,
291, 293, 294, 296

Legitimacy 74, 116, 131, 281, 283,
291, 293
Linear Programming 33, 35
Links 18, 28, 36, 42, 44, 71, 88,
107, 111, 132, 135, 136, 215, 218,
221, 223, 225, 227, 230, 231,
233–289, 296

Major Subsystems 44, 214
Mapping 127, 203, 293, 294
Model 11, 12, 14–16, 25, 29, 30,
34, 41, 42, 44, 47, 56–58, 64, 70,
71, 77, 78, 80, 86–88, 90, 100,
107, 108, 124, 141–145, 147, 153,
154, 185, 220, 241, 243, 255, 264,
284, 286, 297, 299
Monitoring 17, 105, 151, 185,
232, 233, 277, 278

Negotiation 10, 12, 13, 58,
59, 107, 108, 114, 121, 128,
142, 145, 183, 185, 224, 296,
298
NEPA 10, 110, 158, 162, 173,
175, 177, 186, 210, 248, 280,
290
Nested Complexity 19, 20, 26,
43, 215
Neutral 109, 114, 115, 118,
120–122, 131, 135, 136, 141, 185,

188, 195, 197, 241, 242, 245, 246, 280, 293

NIMBY   3, 128, 143, 157

Participation Level Points   112

Permitting Process   110, 159, 175, 177, 180, 184, 185, 188, 189, 205, 210, 245, 248, 262–264, 276, 283, 289, 290

Physical Domain   44, 214

Polarization   60

Policy Lever   91, 94, 219, 221, 225, 230, 276, 287, 288, 296

Positions   1, 6, 68, 118, 121, 123, 125, 126, 129, 140, 142, 171, 176, 180, 189, 193, 196, 207, 245, 248, 252, 253, 258, 284

Power   3, 29, 51, 60, 77, 90, 107, 113, 115, 116, 119, 142, 143, 154, 160, 170, 173–175, 186, 189–195, 197–206, 209–211, 218–220, 223, 227, 241, 243, 247, 256, 259, 260, 291, 293

Pragmatic Analysis   107, 108, 122–126

Process Preparation   109, 185, 186, 241, 279

Public Hearing   1, 2, 111, 150, 159, 177, 210, 245, 248, 280, 283, 284, 290

Public Participation   1, 4, 5, 112, 187, 195, 207

Renewable Energy   159, 160, 170, 174–176, 186, 190, 192–194, 199, 211, 216, 221–223, 235, 236, 259, 260, 280, 281

Representations   17, 20, 42, 69, 71, 77–80, 82, 85, 88, 90–93,

95, 100, 101, 102, 133, 147, 183, 244, 252, 257, 275, 287, 289

Role of Expertise   72, 74

Science-intensive disputes   51, 53–55, 65, 66, 69, 70, 74

Scientific Uncertainty   8, 55, 62, 64

Stakeholder Conflict Assessment   107, 114, 118, 129, 177, 185, 189

Stakeholder Involvement   1, 4–6, 9, 12, 17, 19, 20, 43, 100, 102, 105, 111, 117, 173, 184, 236, 245, 254, 280, 282, 291, 292, 298, 299

Stakeholder Participation   5, 7, 8, 30, 95, 109, 111, 113, 185, 186, 257, 292, 293

Stakeholder-Assisted Modeling   13, 19, 20, 102, 105, 147, 154, 183, 185, 242, 245, 249, 298

Strategic Alternatives   18, 44, 184, 262, 276, 296

System Dynamics   13–16, 40–43, 45–48, 86, 87

Systems Engineering   32, 34, 43, 45, 46, 48

Systems Modeling   13, 25, 34, 36–39, 47, 78, 126

Systems Representation   13, 16, 17, 77, 78, 83, 85, 86, 89, 95, 96, 98, 102, 135, 155, 183, 241, 243, 244, 249, 250, 252, 255, 257, 258, 279, 283, 287, 288, 296

Systems Thinking   13, 47, 106

Timing   184, 257, 280–282

Uncertainty 6, 8, 9, 18, 20, 24,
27–30, 34, 35, 44, 51, 55, 57, 58,
62, 64, 74, 83, 112, 113, 128, 137,
140, 144, 152, 158, 177, 208, 249,
250, 253, 258, 278, 288, 289, 296,
297

Validation 38, 39, 97, 241
Values 8, 11, 20, 29, 33, 35, 38, 42,
43, 46, 47, 51, 56, 58, 60, 61, 66,
70, 77–79, 81–83, 89, 102, 106,

117, 118, 123, 125, 126, 129, 131,
133, 134, 136, 143, 189, 207, 210,
262–264, 270

Wind Energy 1, 2, 33, 154, 157,
159–161, 174, 179, 180, 183, 192,
193, 195, 196, 198, 201, 204, 210,
213, 214, 216, 221–223, 233, 236,
240, 242, 243, 245, 247–249, 258,
260–262, 273, 274, 276, 278,
285, 288, 290, 299